Hitler's Engineers

HITLER'S ENGINEERS

Fritz Todt and Albert Speer –
Master Builders of the Third Reich

Blaine Taylor

CASEMATE
Philadelphia & Newbury

Page 2: Less spectacular than some of Speer's endeavors, but enduring in their own way, were Dr. Todt's sprawling autobahns. A hundred years hence, Speer's reputation may well rest on his wartime ministry and postwar books, while Dr. Fritz Todt will be ranked as one of the world's most renowned road builders. (HHA)

Page 3: Reich Chancellor Adolf Hitler with his two future Armaments Ministers, Drs. Fritz Todt (center, in civilian suit) and Albert Speer (far right, in profile) on a 1936 visit to Düsseldorf to inspect the model of an autobahn bridge. At the time, the much more renowned Dr. Todt was the chief building engineer for the Third Reich, while the younger Dr. Speer was General Building Inspector for Berlin (in 1937). (HHA)

Right: Part of Speer's ethereal show of light and dark over top of a Nuremberg Nazi Party Congress, 1938, which British Ambassador to Berlin Sir Neville Henderson termed "a cathedral of ice" in his wartime memoirs. (HHA)

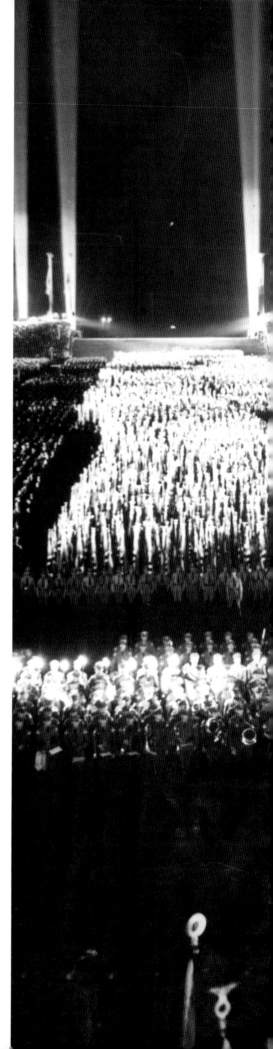

Published in the United States of America and Great Britain in 2010 by
Casemate Publishers
908 Darby Road, Havertown, PA 19083
and
17 Cheap Street, Newbury RG14 5DD

Copyright 2010 © Blaine Taylor

ISBN 978-1-932033-68-7

Cataloging-in-publication data is available from the Library of Congress
and the British Library.

10 9 8 7 6 5 4 3 2 1

Printed and bound in China by Printworks International.

For a complete list of Casemate titles please contact:

CASEMATE PUBLISHERS (US)
Telephone (610) 853-9131, Fax (610) 853-9146
E-mail: casemate@casematepublishing.com

CASEMATE PUBLISHERS (UK)
Telephone (01635) 231091, Fax (01635) 41619
E-mail: casemate-uk@casematepublishing.co.uk

CONTENTS

DEDICATION

To those who helped make this book possible: mentor P. James Kurapka, Gate City, VA; translator Erika Burke, Pearland, TX; and photographer John Durand, Towson, MD, all USA. Such a detailed work as this is truly the sum of all its many parts!

ACKNOWLEDGMENTS

This study is the result of almost five decades of research, but would not have been possible without the valuable help of many people, all of whom I want to mention by name.

First, there is my publisher, David Farnsworth—and my editor, Steven Smith—for agreeing to undertake the project, and for having patience with me while I completed the book after it was due. Just as, without the engineers and architects there would be no roads and no buildings, so, also, without them, there would be no author, and no book. Second, there are the entire photographic staffs at both the Library of Congress in Washington, DC and the United States National Archives at College Park, MD, as well as many other private and museum sources worldwide, credited where their photography and artwork appear.

The photographs would not exist without the excellent pictorial consulting of Stan Piet of Bel Air, MD, and the artistry of John Durand of Towson, MD. Also joining the ranks of those helping me with this project was Renée Klish, Curator of the US Army Combat Art Collection at Fort Lesley J. McNair in Washington, DC. My friend, architect Paul Gorman of Towson, MD, aided in my understanding of various engineering and architectural matters and terms. Lisa Julius of Towson, MD, transcribed the captions of several wartime videos.

As with prior volumes, Mrs. Erika Burke served as my German translator, and both P. James Kurapka of Gate City, VA, and Ray and Josephine Cowderry of North Dakota brought important new Todt and Autobahn photo collections to my attention. Ruth Sheppard in the UK was the copy editor, while Frank White of Towson, MD, and Tara Lichterman of Casemate were computer consultants.

Also, kudos to the wonderful research staff at Baltimore's Enoch Pratt Free Library, and Towson's Baltimore County Library.

My thanks to them all!

AUTHOR'S PREFACE

I first began reading books on Nazi Germany about the time of the 1960 publication of William L. Shirer's *The Rise and Fall of the Third Reich*, which I reviewed later in my Calvert Hall High School literary magazine. This capped my interest in all of the various postwar movies and television documentaries on the war and the Nazis that appeared over the previous decade. Since all of the books that I read left many unanswered questions—written as they were either by observers of the events or authors working with second-hand information—by the time I reached Towson State College in 1968, I'd just begun to do what our history professors urged us to: "Go back to the original sources," the memoirs of the actual participants themselves.

None were more interesting than those of Dr. Albert Speer's, *Inside the Third Reich: Memoirs*, which I reviewed for my college student paper, *Towerlight*. It answered many questions—but not all. Six years later, I reviewed Professor Speer's second book, *Spandau: The Secret Diaries*, for the then *Baltimore News* American newspaper, as I did his third and last postwar published book, *Infiltration: How Heinrich Himmler Schemed to Build an SS Industrial Empire* in 1981.

In those first readings, I was inclined to give Speer the benefit of the doubt, to accept his account of things as the way they actually were; I even wrote to him for an autographed picture—and received it. In the 1990s, I read all three books over again, as well as several others debunking Speer's own accounts. In the preparation for the current volume, I read all of them yet a third time over.

From these multiple readings, I have come to some conclusions. Far from being the apolitical, artistic technician that he claimed to be until the time of his death, Speer was, in fact, a practitioner of the very highest order of Nazi politics and manipulation. He was also an accomplished actor and liar: he knew all about the Holocaust from the very start, as did, indeed, all the other top Nazi leaders. That is why he was treated as such a pariah at both Nuremberg and Spandau by his fellow convicted and accused war criminals: they knew what a phony he was, having observed him in Hitler's presence for almost 15 years.

Thus, the current volume will entirely dispense with all the hand-wringing, back-and-forth "what-did-he-know-and-when-did-he-know-it?" soul-searching of the last few Speer biographies. Based on all of my reading from all sources, and my 63 years of military, business, and political life in a modern, industrialized, technocratic, Western state, I've come to the inescapable conclusion: *he knew*.

That having been said, this book marks a radical departure from all my previously published volumes in one important respect. In them, I baldly stated the facts as I found them, and then let the reader decide. I shall do the same here, but with the following major change: I shall state all sides of an issue, but also give my own considered view. The reader will still be free to make his or her own decision.

I also contest Dr. Speer's assertion that the job of Minister of Armaments found him, rather than the other way round. It is my contention that he actively sought some such high post following the great German military victories of that high tide summer of Nazi victory of 1940, realizing as he did that only through such a top position could he not only maintain his standing within the Nationalsozialistische (National Socialist or NS/Nazi) hierarchy, but actually advance it, especially in the expected postwar Nazi world order that he, and others, felt sure was coming. Therefore, he could not allow himself to remain forever simply Hitler's architect, despite his protestations to the contrary in all his published postwar books.

In addition, I have tried to accurately portray the inner sanctums of Hitler's reign as the principals themselves viewed them; a visual recreation of their world, as seen from within. In this regard, the previously unpublished photographs from their own court photographers' albums are particularly useful.

This is the dual biography of two men, however, not just of one, as well as of their major assistants and—in some cases—rivals. Ironically, the second man—Speer—became more famous only because of the death of the first: the engineer Dr. Fritz Todt, by all accounts also one of the Führer's best and most competent government cabinet ministers.

Over the last 50 years, Dr. Todt has been overshadowed by his successor for a number of reasons: he did not live out the war; he died a mysterious, and as yet unsolved death; he did not stand trial as Speer did at Nuremberg; and he left behind no volumes of memoirs.

In their time, Dr. Todt was far better known than the younger, junior architect Speer: within Nazi Germany, across German-occupied Europe, and around the world as the premier road builder and fortification construction genius of his day.

What began the shift in emphasis was the 1947 publication of English writer Dr. Hugh R. Trevor-Roper's groundbreaking work, *The Last Days of Hitler*, and this trend has continued right up to the present day. It is the intention of this study, therefore, to reverse this trend, to set things aright, to see both Dr. Todt and Dr. Speer in their time as they were, as well as afterward. In such a way, I trust that the scales of objective history will be leveled.

Wherever possible, I have striven to let the players in the great drama speak to us in their own voices at the time, as derived from their public speeches, published writings, and books. This, then, is Nazi Germany as its leaders saw it, not from the prism of those who defeated it.

Finally, this work is a presentation of the political system of Nazism as seen through its very own, favorite lenses: the artistic and scientific disciplines of architecture and engineering, topped off with the joint pursuit of both Drs. Todt and Speer, as well as that of the Third Reich: the manufacture of armaments for war and conquest, all to be completed by 1950 in the new Nazi World Order that remained but an "unachieved empire" in the words of Prof. Alan Bullock.

Blaine Taylor
Berkshires at Town Center
Towson, MD
USA
February 1, 2010

A previously unpublished wartime photo of Dr. Fritz Todt wearing Luftwaffe uniform and flying his own plane. Was he piloting the plane in which he died? We'll never know for sure. (HHA)

CHAPTER ONE

DR. FRITZ TODT: NAZISM'S MASTER ENGINEER-BUILDER

"The master builder who builds in the stone-ocean of a great city must envision his creation amidst the forms and modes of human expression of earlier times. He must express the greatness of our time in relationship to the accomplishments of earlier periods, but the attitude of the master builder who is called upon to create in the wide open space of the German landscape must be altogether different.
"His building site is the wide room of nature. The attempt to be even more monumental, even greater than nature, will seem arrogant and presumptuous."

Fritz Todt in Eduard Schönleben, *Fritz Todt: The Man, the Engineer, the National Socialist*, 1943

Above: Dr. Todt seen reading during the war in a previously unpublished photograph. (HHA)

Opposite: Previously unpublished oil portrait of Dr. Fritz Todt by Wilhelm Otto Pitthan. On his left pocket are his Nazi Party Golden Membership Badge (right), and his World War I medals: Imperial Observer's badge (left), Wound badge (left), and Iron Cross (right). (US Army Combat Art Collection)

Shortly after being named German Reich Chancellor on January 30, 1933, Adolf Hitler announced his intention to build the world's best system of highways—the *Autobahnen* (autobahns)—as well as Europe's first inexpensive common person's car to drive on them.

Hitler's domestic roads were partially completed by the time that his mechanized armies invaded Poland in September 1939, and proved of military value by enabling the armies of Nazi Germany to move both east and west in record time, although the bulk was still handled by the more traditional railway system. In 1945, the Allies would use these same autobahns to overrun Nazi Germany.

"Hitler's highways," as they were called, helped the Nazi Party eliminate unemployment in Germany in a bare six years according to many period scholars, although that long-held conclusion has been disputed in the 2007 book *The Wages of Destruction*. The fact remains, however, that Hitler and his Nazi regime did away with unemployment within six years; a feat unheard of up to that time, and unequaled since in any modern, industrial state anywhere in the world under any political ideology.

Even today, over seven decades later, the autobahns remain engineering marvels of the modern world, and the Nazi-inspired Volkswagen Beetle became the best-selling car in the world just as Hitler predicted. Today, they traverse Central Europe on the Führer's most enduring legacy: the concrete ribbons that encompass the modern German democratic state.

Both Hitler and the Czech-born car genius Dr. Ferdinand Porsche the Elder—designer of the prototype VW—are names well-known in the West, and, indeed, around the world, but not so that of the father of the Autobahnen, Hitler's premier engineer, Dr. Fritz Todt, who also built the German West Wall (known to the Allies as the Siegfried Line), and who started construction of the Atlantic Wall. Who was Fritz Todt, and why was his loss in an airplane crash on February 8, 1942 so lamented by his brother Nazis?

Fritz Todt was born in Pforzheim, Baden, Germany on September 4, 1891, the only child of Emil and Elise Todt. His father was the owner of a jewelry factory. Albert Speer wrote of the similarities between Todt and himself: "We had much in common... Both of us came from prosperous, upper middle class circumstances; both of us were Badeners and had technological backgrounds. We loved nature, life in alpine shelters, ski tours..."[1] He attended public primary and secondary schools in Pforzheim, graduating in 1910 with a "very good" certificate. Following a year's enlistment in the Baden Field Artillery Regiment 14 in Karlsrühe, he studied at a *Technische Hochschule* (college of technology) in Munich and in Karlsrühe from 1911 to 1914.[2] His studies were interrupted by World War I. At Todt's funeral in 1942, Hitler summarized his service in World War I:

> At the outbreak of war, he joined the fourth regiment of field artillery and first saw action on the Western Front. In October 1914, he was named a lieutenant of the reserves, and assigned to the 110th Regiment. With this outfit, he fought up to January 1916. Then he joined the Air Force, became an aerial observer, and was finally leader of an independent flight squadron up to war's end on the Western Front. He was also wounded in an air battle.[3]

From August to October 1914 he served on the Western Front with the Baden Field Artillery. Afterwards he became a lieutenant in the reserves in Baden Infantry Regiment 110. Having joined the Air Force in January 1916, in June he became "leader of the *Reihenbildtruup* (Reihen Picture Troop) of Army Group C, Mars-la-Tour/Briey, France. In August 1918, he was wounded in an air battle."[4] In all, Todt was wounded twice, and was decorated with the Iron Cross 1st and 2nd Class, the Hohenzollern Order, the Bavarian Military Service Order, and the Baden-Zahringer Lion Order.

After World War I, Todt returned to his studies. He completed his professional studies in 1920. He worked first as a laborer, before settling at the Munich civil engineering firm of Sager & Wörner in 1921. He was soon promoted to construction foreman at a water power plant project at Ulm on the Middle Isar River. Beginning in 1925, Todt studied the latest construction methods of the day, leading to his next promotion, when he became Sager & Wörner's technical leader and manager. He then specialized in modern road surfacing, working all over Weimar Republican Germany, from Bavaria to Pommern, from East Prussia to Hanover, from Saxony to

Dr. Todt (left) looking over the shoulder of Nazi Motor Korpsführer (Corps Leader) Adolf Hühnlein during a joint inspection tour of autobahn construction sites in 1935. The two top Nazis worked closely together, just as, later, Dr. Speer would also cooperate with the German Inspector General of Automotives, Jakob Werlin. (HHA)

General Inspector Doctor of Engineering Fritz Todt, builder of the Reich Autobahnen, an official portrait that appeared as the frontispiece of his 1943 biography. (HHA)

On January 12, 1941, Dr. Fritz Todt (left) presented Reich Marshal Hermann Göring (right) with a book on fortress building on Göring's 48th birthday, to his evident pleasure. Dr. Todt was Göring's commander in two separate capacities, both as a Major General in the Luftwaffe, and also as his Four Year Plan appointee for construction projects—an unusual position to be in with one's rival! (HGA)

Württemberg, from the Saar to Pfalz. As the later Nazi-made film *Dr. Todt: Man, Mission and Achievement* asserted in 1943, "Times were hard. Prospects were poor for engineers."

Todt joined the Nationalsozialistische Deutsche Arbeiterpartei (German National Socialist Workers' Party or NSDAP) on January 5, 1922. In 1931, he was made an Schutzstaffel (Protection Squad, or SS) Standartenführer (Colonel) on the staff of Heinrich Himmler. Hitler's eulogy summarizes his progression within the Party:

> In 1931, he joined the SA [Sturmabteilung (Storm Detachment)] ... starting as an ordinary Storm Trooper. He then became a squad leader, in the same year was advanced to standard bearer, and by 1938, had risen to Chief Leader, Brigade Leader, and—finally—Chief Brigade Leader, and was active as well in the Party ... an associate of the progressive League of German Architects and Engineers in Munich ... and Technical Consultant of Highway Construction in the Office for Economic Coordination and Work Procurement of the Party.[5]

In 1931, Todt completed his doctorate. His doctoral dissertation was a paper based on his personal experiences at construction sites and his specialization in modern road surfacing, entitled *The Causes of Defects of*

DR. ROBERT LEY AND THE DEUTSCHE ARBEITSFRONT (DAF)

Doctor Robert Ley's early, and enduring, support of Adolf Hitler meant that he achieved great power despite the negative opinions held of him by most of the rest of the Nazi leadership. In World War I his aircraft was shot down, and he became a prisoner of war. He almost lost a leg, and sustained frontal-lobe damage which affected his mental condition, leaving him with a stammer, and increasingly, an alcohol addiction. (A car accident in 1930 may have exacerbated the damage.) However, once released from captivity, he renewed his studies, and obtained his doctorate in chemistry within six months.

He gave up his job to devote all his time to Party affairs in the mid-1920s, a risky move that showed his complete conviction, which Hitler never forgot. He rose within the Party, becoming Reich Organizational Director in 1933. In May 1933, he helped Hitler solve one of the thorniest problems facing the new Nazi regime: what to do about the massive labor movement of the Weimar Republic? The simple solution adopted was to outlaw it, then seize all its offices and assets.

A few days later, the Labor Front (DAF) was created by constitutive congress. The DAF replaced all free trade unions, being an organization that represented all workers. Though technically voluntary, membership was "desired," and in 1942 it had 25 million members, making it the largest labor organization in the world. Members paid compulsory dues from their wages. When Germany went to war, the DAF was an essential participant in converting the economy to war production. Dr. Ley was Göring's subordinate within the German Four Year Plan economic organization, and the portly Field Marshal ordered him about "like a waiter," according to one observer.

Ley's leadership of the DAF made him one of the most domestically powerful men in the new Nazi Germany. He is best known for being the

Above: A previously unpublished official portrait of Dr. Robert Ley, head of the German Labor Front (DAF) and *Gauleiter* (District Leader) by wartime combat artist Pitthan. On his breast pocket he wears, top to bottom, beneath the row of ribbons, the Hitler Youth Golden Badge of Honor, the Golden Party (Membership) Badge (right), the World War I Imperial Observers Badge (left) and (below right) the World War I Wound Badge. Note also Dr. Ley's Nazi Party Gauleiter's armband. (Courtesy Mary Lou Gjernes, former Curator, US Army Combat Art Collection, Fort Leslie J. McNair, Washington, DC)

Left: A prewar photo of, left to right, German Army Field Marshal Werner von Blomberg, three unidentified persons, DAF Leader Dr. Robert Ley, Army Gen. Wilhelm Keitel, then von Blomberg's deputy, unknown people, and Army C-in-C Col. Gen. Werner von Fritsch. (Photo courtesy John F. Bloecher, Jr., Danzig Report, LC)

On May 1, 1933—"May Day" in Socialist and Communist countries, celebrating labor—the Nazis carried out their long planned takeover and dismantling of all independent trade unions and the Sozialdemolcratische Partei Deutschlands (Social Democratic Party of Germany or SPD) in Germany. Here, Dr. Ley (third from right) and his staff meet during April to make their final plans. Left to right: Brinckmann, Biallas, Peppler, Schmeer, Müller, Ley, Schümann, and Muchow. (HHA)

head of the sub-organization "Strength through Joy" that provided German workers with subsidized vacations and leisure activities. Dr. Ley supported a minimum wage and better worker housing, and two luxury cruise ships for worker vacations abroad were launched under his regime. The "Beauty of Work" was another sub-organization which aimed to make workplaces more appealing to workers. The Reicharbeitsdienst (Reich Labor Service or RAD), was another arm of the DAF.

He was at times both an ally and a rival to Drs. Todt and Speer, the war bringing him into conflict with both of them. Captured by the 101st Airborne Division in 1945, he committed suicide in prison while awaiting trial at Nuremberg. After Ley's suicide, Göring said "It's just as well that he's dead... I'm sure he would have made a spectacle of himself at the trial."

Dr. Ley (in civilian clothes, second from right) gives a backhanded and modified Nazi salute (not a wave) aboard the "Strength through Joy" (KdF) ship named after him, as happy vacationers join in the merriment. The ship was christened by Hitler, who in 1939 joined the Leys for the "only vacation of his life" aboard it. (HHA)

Dr. Todt (left) chats with Gen. Wilhelm Keitel—Chief of the Oberkommando der Wehrmacht (High Command of the Armed Forces or OKW,) the High Command of the Armed Forces—on March 1, 1939, the "Day of the Luftwaffe," in the Great Hall of the Air Ministry Building in Berlin. Having helped Hitler end German unemployment during the first six years of the Nazi regime (1933–39), Dr. Todt also aided the Führer in preparing for war by constructing the West Wall (Siegfried Line) against France. Appointed Minister of Armaments and Munitions on March 17, 1940, as well as Inspector General of Water and Energy on August 6, 1941, Reich Minister Dr. Todt's wartime building projects included the concrete U-boat pens on the northern coast of France and the start of construction of the Atlantic Wall. Here, standing between the two men, is Nazi Minister of Education Dr. Bernhard Rust, while at far right is another Todt rival, Reichsführer-SS Heinrich Himmler. (HGA)

Asphalt (Blacktop) on Roads (also known as *Reasons for Paving Mistakes of Roads Paved with Tar and Asphalt*). Hitler's eulogy continued:

In 1932, [his activities] resulted in the NS German Technical Union coming under his leadership... In connection with the opening of the Automobile Exposition in 1933, I tried to realize ... improvement of the German road network already in existence, but also in the field of construction of new, special auto roads.

This was a general plan which essentially only embraced the general principles. In Dr. Todt ... I believed I had found a man who was suited to transform a theoretical intention into a practical reality. A brochure published by him about new ways of road construction was submitted to me, and especially strengthened me in this hope.

A previously unpublished photograph of Hitler's daily walk from The Berghof to the Mooslahnerkopf Teahouse. Following tea and cakes, the walkers were then driven in waiting cars back to The Berghof. On these walks, the Führer gave the place of honor at his right to the day's special guest, in this case Dr. Todt. Hitler wears soft felt fedora and early SA jacket. (EBHA)

NAZI CONSTRUCTIONS (PARTIAL LIST), 1933–45

Anhalter Bahnhof/Train Station

Atlantikwall/Atlantic Wall

Autobahnen/Highways

Berghof/Mountain Home, Obersalzberg, Bavaria

Braunes Haus/Brown House, Munich

Congress Hall, Nuremberg

Concentration camps. Germany and Nazi-occupied Europe

Deutches Stadion/German Stadium, Nuremberg

Ehrentemple/Honor Temples, Munich

Flakturm/Flak Tower, Germany

Frankischer Hof/Frankish Home

Führerbau/Leader Building, Munich

Führerbunker/Leader Bunker, Berlin

Führerhauptquartiere/Leader Headquarters, Germany, Austria, France, Belgium, Ukraine

Gaubunker/Regional Bunker

Gauhaus/Regional House

German Air Ministry, Berlin

Hall of Models, Berlin

Haus der Deutschen Kunst/House of German Art, Munich

Hitler Jugend Heim/Hitler Youth Home

Jena Brucke/Bridge

Jugendherberge/Youth Hostels

Konigsplatz/King's Plaza, Munich

Kehlsteinhaus/Eagle's Nest on Kehlstein Mountain, Bavaria

Nazi War Memorials, Berlin, Munich, Nuremberg

Nazi Party Congress Rally grounds, Nuremberg

Obersalzberg/Over the Salt Mine Mountain, Bavaria

Olympic Stadium, Berlin

Prora, Rugen Island

Neue Reichskanzlei/New Reich Chancellery, Berlin

Soldatenhalle/Soldiers' Hall, Berlin

Tempelhof International Airport, Berlin

Thingplatz/Thing Plaza or Thingstatte

Triumphal Arch, Berlin

Volkshalle/People's Hall

Winkelturme/Winkel Tower

Zeppelinfeld/Zeppelin Field, Nuremberg

After long discussions, I entrusted him on June 30, 1933 with the task of building the new Reich's auto roads, and ... the general reform of the whole German highway construction system, as General Director for the German Highway Construction System...

During the next decade, the modest, unassuming technologist gathered into his hands responsibility for the entire German construction industry ... in charge of all navigational waterways and power plants.[6]

As Speer said in his memoirs: "Hitler, too, paid him and his accomplishments a respect bordering on reverence. Nevertheless, Todt maintained his personal independence in his relations with Hitler, although he was a loyal Party member of the early years."[7] He also said of Todt:

Dr. Todt was one of the very few modest, unassertive personalities in the government, a man you could rely on, and who steered clear of all the intrigues. With his combination of sensitivity and matter-of-factness, such as is frequently found in technicians, he fitted rather poorly into the governing class of the National Socialist state. He lived a quiet, withdrawn life, having no personal contacts with Party circles—and even very rarely

The Reich autobahn marker for the route from Munich, Germany to Salzburg, Austria serves as a backdrop for this prewar Mercedes motorist, German race car driver Rudolf Caracciola. (Previously unpublished photo from the Heinrich Hoffmann Albums, US National Archives, College Park, MD)

A previously unpublished prewar photo of Hitler (left) with Dr. Todt (right) at an autobahn opening. (HHA)

appeared at Hitler's dinners and suppers, although he would have been welcome. This retiring attitude enhanced his prestige: whenever he did appear, he became the center of interest.[8]

The editor of the partially postwar-published Goebbels diaries, American journalist Louis P. Lochner, wrote that, "Todt ... was one of the top Nazis whom foreigners liked. He had none of the strutting, arrogant mannerisms of the typical Party brass hats. Modest and unassuming, he made many a foreign visitor wonder how he ever got mixed up with Nazism."[9] Despite holding the rank of General in the SA, SS, and Luftwaffe (Air Force), Todt's reputation emerged unscathed by the Nazi crimes of World War II, unlike that of his successor, Speer. As Matthias Schmidt noted:

> Todt had devoted much thought to the Autobahn project, which he saw as a means of ending unemployment. He was also attracted to the technical aspects of the task. "The plan to build a network of connecting highways," he said, "offered challenges which the master builders of many centuries had longed for in vain... The autobahns must not be alien entities, they have to be integral parts of the landscape."
>
> This and other functions had made Fritz Todt a central figure in the leadership of the Reich, yet he never got entangled in the omnipresent internal intrigues of the Party. His personal modesty, his lack of the bragging and blustering typical of other Party functionaries, gained him the respect of the general population, if only because he never touted his own virtues.
>
> His technological expertise, his dislike of unnecessary bureaucracy, and his lack of interest in Party infighting brought him, as Speer noted, Hitler's "respect bordering on reverence." This was entirely justified by Todt's accomplishments.[10]

It is time to take a closer look at those stellar accomplishments that so endeared Dr. Fritz Todt to his Führer and the rest of the Nazi leadership corps, starting with the Autobahnen.

"UNDER SHOVEL AND RIFLE": THE REICHSARBEITSDIENST (RAD)

In 1931, Reich Chancellor Heinrich Brüning established the Freiwilliger Arbeitsdienst (Volunteer Labor Service or FAD), an employment program for young, unemployed men. When the Nazis came to office, they wanted to distance themselves from this program, so instead established their own Reichsarbeitsdienst (Reich Labor Service or RAD) under former German Army Colonel Konstantin Hierl,[11] in 1933. The RAD became a main supportive organization of the various projects of both Drs. Todt and Speer both before and during the war.

In 1937, a booklet entitled *German Labor Service* by Fritz Edel detailed the RAD's genesis and purpose:

> Sons of miners, civil servants, professors, and farmers work together—shoulder to shoulder—and thus learn the practical significance of the words Nation and Socialism. And no matter to what position in life they return, they bring with them a clear consciousness of the truth that work is not only a means of earning money, but is the moral basis of national life.
>
> By working in the RAD, the youth of the nation is brought to realize the fact that work is a noble thing no matter what form it takes.
>
> The second task before the RAD is to free Germany from the necessity of importing food supplies. When the program allotted to the RAD for the next 20 years is fully carried out, Germany will have gained a new province, represented by reclaimed land.

A previously unpublished view of Colonel Hierl (left) and Hitler (right) facing the newly announced RAD from the Zeppelin Field speaker's rostrum, Thursday, September 6, 1934. Note, too, the telephone and wiring at lower right. Today, the platform remains, but the railing, carpeting, and telephone are all long gone. According to the 1937 Nazi publication *German Labor Service* by Fritz Edel, "The barriers which have divided class from class and creed from creed are excluded from the comradeship of the Labor Service." (HHA)

And all this will be achieved through peaceful effort. Therewith, Germany will be assured of sufficient home produce to feed her whole population...

Colonel Hierl wrote, "RAD signifies something different, something greater than a temporary measure arising from the distress of the time for the purpose of combating unemployment. The idea of compulsory labor service is a logical development and fulfillment of the idea embodied in compulsory education and national military service.

"Every German must work for his country and fight for ... his country. Compulsory labor service must become a duty of honor for German youth in the service of the nation. Its purpose must not be to supply cheap labor for private enterprise, and it must not become a competitive undertaking carried on by the State for the purpose of forcing down the level of wages.

"Through the ... RAD, the national government will have at its disposal a working army that will carry out great public works to serve the economic interests of the nation, as well as its cultural and other public interest."

Colonel Konstantin Hierl as head of the RAD in this previously unpublished formal prewar painting by noted artist Wilhelm Otto Pitthan, complete with ceremonial dagger—so beloved of Nazi uniform designers—at his side. (US Army Combat Art Collection, Fort Leslie J. McNair, Washington, DC)

The key man in the overall saga of the new RAD was Konstantin Hierl. Born in 1875, he joined the Royal Bavarian Army after high school. From 1903, he served as a lieutenant on the Bavarian General Staff, transferring four years later to the Great Prussian General Staff. From 1911 he taught military history at the Bavarian Staff College for three years. He served in the World War I, as Chief of the General Staff of the First Bavarian Reserve Corps, and later as a battalion commander. In 1918, Hierl set up his own Free Corps to combat German Communist Spartacist troops after their seizure of Munich. In April 1919, he reclaimed Augsburg from the Spartacists. He then secured a position with the Reichswehr (Republican Army), where he worked as a liaison officer to the Reich government in Berlin. In 1922, he was promoted to the rank of colonel and assigned to the Reich Ministry of Defense.

In 1923, Hierl sent a memorandum to his superior, the Chief of the Army Command, Lt. Gen. Hans von Seeckt, outlining his ideas on labor service. In this memo, he described conditions affecting young people as he saw them, and recommended basically what later became the Nazi RAD: "Overcome the class system, Marxism, and democracy." Von Seeckt, however, ignored his prescient memo, and as a result Hierl left the Army, retiring on September 30, 1924, after 30 years of service.

Meanwhile, he had met three paragons of right-wing German nationalists: former Imperial Army Quartermaster-General Erich Ludendorff, Nazi politician Gregor Strasser, and Hitler himself, in 1920, when he had attended the Party's first mass rally that February 24.

He hadn't joined right away, he later asserted, because—as a serving officer—he wasn't allowed to. The retired Army colonel was also known as a military writer who fancied the coming doctrine of total war that would be avidly embraced during the future Third Reich by Hitler, Dr. Josef Goebbels—and Professor Albert Speer in 1943.

Thus it was that Colonel Hierl, 54, joined the Party on Hitler's 40th birthday, April 20, 1929, just six months before the onset of the Great Depression that would elevate the Nazis into national office and create for Hierl the future mass membership of his RAD.

In 1930, Colonel Hierl was elected to the Reichstag (Parliament) as a Nazi deputy, and also sent to his Führer his old memo on labor.

In June 1932, Hitler appointed the retired colonel as Commissioner of the Führer of the Nazi Party for the Labor Service to rival the Weimar Republican FAD. In *Soldiers of Labor*, Professor Patel asserted: "His mania for planning went so far that he devised a food plan reckoned down to grams and pennies, and regulated even the number of kerchiefs and drinking cups... The Nazis based their concept of the labor service on military organization."[12]

Also in 1932, Hierl loosely aligned his projected RAD with a similar organizational set-up of the German World War I veterans association, the Stahlhelm (Steel Helmet), led by Franz Seldte who, on February 2, 1933, became Reich Labor Minister in Hitler's first "Cabinet of Barons" (1933–37). Named as Seldte's deputy, Colonel Hierl began paramilitary training of the future RAD right away.

The former wartime Allied powers briefly halted the growth of the new unit because it seemed too closely akin to a civilian militia—aligned with the real military—and thus forbidden under the terms of the Treaty of Versailles of 1919.

The RAD was not a big hit with the German public at first during 1933–34, and was even almost placed under SA Chief of Staff Capt. Ernst Röhm—who had both the nascent SS and Nationalsozialistisches Kraftfahrkorps (National Socialist Drivers' Corps or NSKK) in similar subordinate capacities—in 1934; but it wasn't, and thus Colonel Hierl and the RAD survived the June 30–July 2, 1934 "Blood Purge" against Röhm and his top SA leadership cadre officers.

Indeed, on July 3, Colonel Hierl's RAD found itself removed from the Reich Labor Ministry, and reassigned to that of Nazi jurist Dr. Wilhelm Frick, Reich Minister of the Interior, instead. Two months later—on Thursday, September 6, 1934—the RAD (along with the black SS of Reichsführer-SS Heinrich Himmler) became one of the star organizations given gala onscreen treatment in German director Leni Riefenstahl's epic and epochal film *Triumpf des Willens (Triumph of the Will)* that instantly made it known worldwide, especially across the Third Reich and Europe.

In these cinematic scenes, Nazism's heroic-looking "Soldiers of Labor"—with their shiny, polished spades, peaked caps, snappy uniforms, and handsome physiques—made all the German girls swoon with excitement, and German men want to enlist right away. The RAD had finally arrived, a decade after its initial conception by an unknown Army colonel from Munich.

Left: The RAD on the march at Nuremberg, September 1934, as seen in this previously unpublished photograph. (HHA)

Below: Wartime RAD men at work in the occupied USSR. Among their duties on the Eastern Front, the RAD men guarded prisoners of war, built ghettoes, fought Red Partisans, and took part in "racial cleansing" operations against Jews, gypsies, and others, and yet Colonel Hierl wasn't indicted by the International Military Tribunal (IMT) at Nuremberg after the war as a war criminal. (CER)

The RAD was legislatively codified by law on March 26, 1935 as a compulsory Nazi organization. Hierl was promoted from colonel to major general in May 1936, and the following September also named a Party *Reichsleiter* (Reich Leader).

In March 1938, Austria was added to Hierl's domain, followed in October by the Sudetenland. By summer 1941, the RAD's labor pool empire encompassed Norway, the Netherlands, Flanders, Wallonia, Croatia, and fascist Slovakia as well, all of which had RAD branches on their soil.

When Hierl turned 70 in 1945, Reichsführer-SS Himmler replaced Dr. Frick as Reich Minister of the Interior, thus making the RAD a top-level Reich agency at last. In the spring of 1945, Hitler prevented Himmler from absorbing the RAD outright into his still-growing SS industrial empire.

Originally, the RAD's Berlin headquarters had been in the Reich Labor Ministry building, then moved to its own location at Berlin-Grunewald, near the famous city forest of that name, in 1936. The RAD's membership figure peaked at 340,000. On April 1, 1936, the former autonomous organization known as the Women's Labor Service was incorporated into the RAD. It was compulsory for all German men to serve for a period up to a year in the RAD at some point between the ages of 18 and 25. They usually served before beginning their university or military careers.

RAD members provided service for various civic, military, and agricultural construction projects. Initially they worked on drainage works, constructing dikes and irrigation channels, and cultivating land, all with the aim of providing new farmland for the Reich. They planted forests, repaired river banks, and reclaimed wasteland, Although the RAD played no role in the building of Dr. Todt's autobahns, it did help build the West Wall in 1938–39. During the war their work included building concentration camps, manning antiaircraft guns, and defending the Atlantic Wall.

In 1937, it was claimed that in three years the RAD had reclaimed 300,000 acres of marsh and swamp; improved 750,000 acres of watery soil through work on waterways and drainage; made a further 60,000 acres suitable for arable farming; redistributed around 100,000 acres of land to "give profitable results," and improved accessibility to 400,000 acres of cultivated land. Edel concluded that "The increase in agricultural produce resulting from this work has reached the annual value of 50 million Reich Marks, which is equal to the whole produce of a district as large as that of the Saar [on the border between Nazi Germany and Republican France]."

On Hitler's 56th birthday—April 20, 1945—the Allies formally dissolved the RAD. Hierl was later convicted by a West German de-Nazification court and sentenced to a prison term of five years, but did not serve it.

CHAPTER TWO

"THE FÜHRER'S ROADS": NAZI AUTOBAHNEN RIBBONS OF CONCRETE, 1933–45

"German workers, to work!"
Adolf Hitler, Frankfurt am Main, September 23, 1933

A good overview of an *Alpenstrasse* (Alpine Highway) bridge, from *Germany's Highways: Adolf Hitler's Roads*, 1937. Other structures included the Werra Valley Bridge near Hannover-Münden on the Göttingen–Kassel Autobahn route that was built between October 1935 and April 1937, with steel construction added between July 1936 and January 1937. (LC)

On March 28, 1938, there appeared in the famous American picture magazine, *Life*, the following photo caption:

Action and talk are the twin screws of fascism. In Germany, the endless rain of talk is supplied by [Nazi Propaganda Minister Dr. Josef] Goebbels. But, toward the day of action, the German Army has criss-crossed Germany with the world's most magnificent skein of roads, the Reichsautobahnen … [made up of] two 25-ft tracks, divided by 16 ft of lawn, with no crossings.

Four cars can run abreast on each section. There is no speed limit; 600 miles have thus far been finished by the German unemployed. These head straight to the borders of Austria, Czechoslovakia, Poland, France, Belgium, the Fatherland, and to the Baltic.

In case of another World War, Germany would be fighting once again on interior lines, able to move armies quickly from front to front.[1]

The Pre-Autobahn Era

The idea of limited-access automobile highways dates back to the New York City Parkway system, the building of which began in 1907–08. On January 23, 1909, the Automobil-Verkehrs-und-Unbungsstrasse GmbH (AVUS) opened in Berlin, then capital of the Imperial Hohenzollern dynasty of the Second German Reich of Kaiser Wilhelm II. A private venture, AVUS GmbH was financially supported by racing and business interests, and was designed and engineered as both a racing and test track, consisting of a pair of 8m-wide lanes separated by a 9m median strip. This was followed in 1912 by an "intersection-

free" AVUS, also in Berlin, that ran 9.8km from the former Halensee-Nikolaisee rail line to the Teltow Canal in Charlottenburg. With construction interrupted by the outbreak of World War I in 1914, this new AVUS extension wasn't completed until 1921 and, again, was essentially a closed racing and test track once more.

That same year, however, industrial tycoon Hugo Stinnes bought the private roadway and expanded it again, this time to four lanes running 1.5km. Later, AVUS became part of the overall Berlin municipal public road network, and today is part of the city's A115 autobahn.

In 1924, Italian Prime Minister Benito Mussolini attracted the attention of Adolf Hitler when he opened the world's first limited-access expressway, the 130km (80 mile) long Autostrada dei Laghi (Lakes' Highway) from Milan to Lake Como in northern Italy, Fascism's birthplace. Similar to Germany's autobahns, the autostrada actually pre-dated Fascism, having been started on September 21, 1921, but Mussolini took credit for it, a tactic employed by his German imitator in 1933. Unlike the later Nazi autobahn, however, the autostrada was a toll road, and didn't provide divided lanes until later. The Duce's friend and engineer Piero Puricelli designed the Milan–Varese section of the autostrada. Puricelli covered the expense of the autostrada by introducing a toll to be paid by drivers on the new motorway.

Following the success of the autostrada, the Weimar Republican German Studiengesellschaft für den Automobilstrassenbau (or Stufa for short) began planning and theoretical research for a domestic highway network similar to that of their Italian neighbor. As Arend Vosselmann's German-language study *Reichs-Autobahn: Schönheit— Nature—Technik* (*The Reichsutobahn: Beauty, Nature, and*

Technology) notes: "In 1924 Germany, the road system was in unorganized chaos, with each state government having responsibility for its own roads. A central agency was without influence, and therefore useless. The growing auto traffic with its high speeds made it necessary to develop new roads by way of a unified plan."[2]

In 1926, Stufa published the *Vorentwurf zu einem Kraftwagebstrassennetz Deutschlands*, an ambitious plan for a 22,500-km (13,980-mile) German superhighway network. (Today, the modern autobahn network is still but half that comprehensive length.) This would be the first national freeway system. The year 1924 also saw Weimar Germany's first traffic signal go into service at Berlin's Potsdamer Platz (Plaza), the city's busiest intersection.

On November 6, 1926—under the leadership of Willy Hof, the chairman of the German Chamber of Commerce—there was founded the Verein zur Vorbereitung der Autostrasse Hansestadte–Frankfurt–Basel (Association for the Planning of the Hanseatic Cities–Frankfurt–Basel Motorway), shortened to the acronym HaFraBa. The HaFraBa planned and researched the future building of Germany's first modern highway system.

Although HaFraBa initially planned to finance its "cars only" highways with toll charges plus rest stop restaurant and gas station income, German law subsequently prevented any such tolls being placed on the new motorways. In contrast, the modern autobahns of today are financed by taxes on gasoline and oil, while there is also a German autobahn toll system for trucks.

"The cuts into the land are enormous! Tree trunks are dynamited, but the treasured earth was carefully saved in compost heaps. It will be used later for plantings in median strips and embankments, or in forming areas." Note the rail tracks and trio of locomotives. (*OT/LC*)

AUTOBAHN CHRONOLOGY

May 1, 1933	Hitler announces formal plans to build the Autobahnen
June 27, 1933	Passage of Reichsautobahn (RAB) Law by Cabinet
June 30, 1933	Hitler appoints Dr. Todt as General Inspector of the German Road System
September 23, 1933	First groundbreaking at Frankfurt am Main
June 9, 1934	Founding of publication *Die Strasse* (*The Roads*) in Munich
September 3–19, 1934	Seventh International Road Congress in Munich and Berlin and International Road Building Exhibition
September 7, 1934	Dr. Todt's Autobahn Report to the Nazi Reich Party Congress at Nuremberg
September 14–November 1934	Zeppelin flights over autobahns for aerial photography
November 7–15, 1934	The Führer tours autobahn construction sites from Munich to the Austrian frontier
February 18–19, 1935	International Automobile and Motorcycle Show in Berlin where *Autobahn* exhibit shown
May 19, 1935	Opening of the first autobahn section Frankfurt am Main–Darmstadt
January 19, 1936	Reich Road used for the Winter Olympics at Garmisch-Partenkirchen in Bavaria
February 15, 1936	International Automobile and Motorcycle Show in Berlin
April 4, 1936	Minister-President of Prussia Hermann Göring opens the RAB section Berlin–Joachimstal
April 25, 1936	Opening of the RAB stretch Halle–Leipzig (26km)
May 21, 1936	Opening of the RAB stretch Cologne–Düsseldorf (25km) and route to Belgium and Holland by Dr. Goebbels
May 22, 1936	Dedication of the Admiral Graf Spee Bridge over the Rhine River between Duisburg and Rheinhausen by Dr. Goebbels
May 24, 1936	Opening of the RAB stretch Wetarn–Rosenhiem (33km) by Dr. Todt
June 7, 1936	Dedication of the Adolf Hitler Rhine Bridge in Krefeld–Uerdingen by Deputy Führer Reich Minister Rudolf Hess
June 19, 1936	Opening of the RAB stretch Königsberg–Kobbelbude (15km)
August 1936	Building of the German Alpine Road section Inzell–Mauthausel–Schwarzbachwacht
August 17, 1936	Opening of the RAB stretches Weissenfels–Eisenberg, Rosenheim–Siegsdorf, Berlin–Magdeburg, Helimstedt–Braunschweig (200km)
September 27, 1936	Opening of the first 1,000km of the Autobahn
October 1–10, 1936	International Congress for Bridge Building and High Bridges in Berlin and Munich under Dr. Todt
October 24, 1936	Opening of the show *The Roads of Adolf Hitler in Culture* in Berlin's Castle Niederschönhausen
November 21, 1936	Opening of the RAB stretch Schkeuditz–Weissenfels (15km)
December 12, 1936	Opening of the RAB stretch Düsseldorf North–Oberhausen (17km)
December 19, 1936	Opening of the section Eisenberg–Schleiz (38km)
December 17, 1937	Opening of 2,000km of the RAB
December 15, 1938	Opening of 3,000km of the RAB

Above: "Manpower alone is not enough to move and transport earth and rocks. Giant cranes are also put to work… As far as rerouting paths and roads is concerned, sometimes the removal of homes is necessary. These measures had to be settled, and the people compensated… Bifurcations and crossings of all sorts had to be made both clear and practical, the right curving and overpasses calculated. In mountainous terrain, rises must be taken into consideration," stated *Germany's Highways.* (OT/LC)

Right: "Earth for the roadbed was often moved by water… After the application of stone, etc., rollers were needed to pack it all down tight and firm. Many cement mixers were used. They required equipment to measure ingredients such as water and sand. Machines were also used to put down an even-layered pavement, and also make it watertight. The concrete is laid in two layers. The first layer needed to be compressed, while the final layer needed to be joined evenly and straight… Much other equipment was used as well: electric pumps for water, machines to pump in the joint material, joint cutters, machines to smooth the cement surface, cranes to lay tracks, conveyor belts, and water sprinkling trucks to keep cement damp," stated *Germany's Highways.* (Archiv, G.I./OT/LC)

That same watershed year of 1926 also saw the landmark merger of the two German automobile production giants, Daimler and Benz, as planning began for a new, limited-access highway to run between Aachen and Cologne. When the worldwide Great Depression hit three years later, construction was limited to a shorter stretch of 20km between Bonn and Cologne. The roadwork began in October 1929 as the Great Depression hit America. In an effort to create jobs for people, initially very few machines were used in favor of human labor, the very tactic adopted in 1933 by Dr. Todt. The road was finished in 1932, the year before Hitler took office as Reich Chancellor of a coalition government of Nazis and Conservatives. Cologne Mayor Konrad Adenauer opened the new autobahn on August 6, asserting that, "This is how the roads of the future will look!" Today, this stretch of road is part of the A555 autobahn, validating Mayor Adenauer's 1932 prediction.

In 1927, a plan was drawn up for the construction of a 22,000-km (13,670-mile) highway network throughout Germany. The idea was filed due to the 1929 New York stock market crash, and the economic crisis that followed.

Even the very first usage of the word "autobahn" was introduced five years before the Nazi Party took office, in 1928, by HaFraBa's public relations director, Kurt Kaftan, who not only invented the term, but also used it as the title of the organization's official magazine.

On July 4, 1930, led by the Nazi Party's majority, the German Reichstag voted down HaFraBa's plans for the Weimar Republican Autobahn national road network. The following year, the First International Autobahn Conference was convened in Geneva to consider a Europe-wide superhighway system with two main goals: the creation of jobs and the promotion of international friendship. The conference was sponsored by the International Employment Office in Geneva, whose director was Albert Thomas.

By 1939, it was Nazi Germany that built the major autobahns in the Third Reich, and by 1941 it was its New Order that was sponsoring a Continent-wide autobahn system that would link Nazi Europe's Atlantic coast in the west to the former Soviet Ural Mountains in East Asia.

Adolf Hitler's Highways

HaFraBa's plans for a countrywide network of free, limited-access, multi-lane highways were at first opposed by the Nazis as being an elitist fantasy, since the nation had so few motorists and cars. In 1933, however, the Führer shrewdly reversed political course entirely, claiming the concept of national autobahns as "Adolf Hitler's Highways."

It was said that as far back as 1924, when Hitler was imprisoned for treason at Landsberg Fortress on the Lech River in Bavaria, he was already interested both in car and national highway development. As the 1937 Nazi book *German Highways* describes it:

> [While in] Landsberg Fortress, Hitler spoke of the necessity that roads must be built to accommodate the technical capability of the motor vehicle, and also to connect all regions. He declared during the time of his captivity … that he would have such roads built someday.
>
> In his 14 years of political struggle for Germany, the Führer used almost exclusively motor vehicles during his travels, and got to know German roads south, east, north, and west as almost no other man did, traveling an estimated 500,000–600,000km [310,694–372,832 miles], thus having traveled at least a dozen times around the world.
>
> In his travels, Hitler had the opportunity to learn about road networks, and out of this knowledge came the idea to build a connecting net of roads…[3]

Hitler became Chancellor of the German Reich on January 30, 1933. On February 11, at the opening of the International Automobile and Motorcycle Show in Berlin, the Führer announced both the new regime's automobile and road-building policies in his address. He noted that, "As in earlier times, when roads had been made for horse and wagon, and tracks laid for railroads, now roads must be built for automobiles."

As an expert on road construction, Todt had prepared many memoranda on the development of traffic systems which also covered financial issues, and that of unemployment. The most famous of these

"Nowhere is manpower replaced by machines. Although not many people *are* seen in the spread-out terrain, they are there… Geologists and earth scientists were needed. The geologist works in the field, takes stone samples, and drills the earth for samples. In the laboratory of the Institute for Earth Science at universities, the results were compared and studied. Scientific experiments and exploration of changes in soil through mechanical tests and weather were performed. The first practical work was the removal of the topsoil. The soil is carefully composted and saved for later use, after the completion of the road, for plantings in median strips and embankments, or donated for farming," stated *Germany's Highways*. (*Archiv, G.I./OT/LC*)

"Busy hands at work… Boulders and rocks had to be moved. The answer to some of the problems was drilling and blasting. Large swamps, wetlands, and bogs presented another problem, and needed to be moved aside. With the help of special machines, they can be lifted out, or cut, and then suctioned dry. Through smart calculations and blasting, room can be made to carry traffic. The movement of soil of the RAB was an enormous undertaking. In 1933, no less than 120,000 workers were employed with spades and machinery." (*Archiv, G.I./OT/LC*)

"Large cement mixers at work. Concrete or coarse mortar—a mixture of sand, gravel, stone chips, bound with cement—was poured while still damp into forms, and allowed to harden… The use of cement on the autobahn in 1936 was about 13% of the total used. Surprisingly, even unbelievably, fast, the German building and machine industry adapted to the need of the autobahn construction, and how much it was needed… On the geological map, hard rock is available in Schleswig-Holstein, Mecklenburg, Pommern, parts of Brandenburg, Posen, and West and East Prussia. The rock lends itself to road construction, rail, and hydraulic engineering," stated *Germany's Highways*. (*Archiv/OT/LC*)

"The autobahn is constructed over a highway. Scaffolding is being built where the cement pillars are erected… Before starting road construction, materials have to be available. After blasting rock, it has to be broken up, and various machinery is needed for that. At the construction sites, earth needs to be moved by conveyor belt, and other equipment is used to smooth the soil, then trucks are used to both bring and remove materials… Rock quarries delivered gravel and rocks for building bridges, plus stone chips, which were used before the final pavement was laid. Sand and gravel pits were reopened," stated *Germany's Highways*. (*Archiv/OT/LC*)

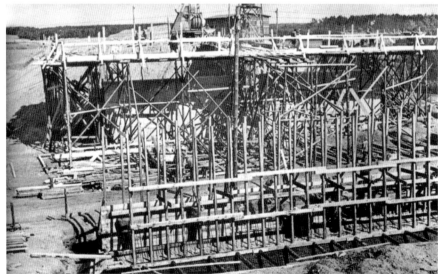

"Pouring the stone layer. (*Archiv, G.I./OT*/LC)

Far left: "Untiring, the large crane lifts the building material from the edge of the river onto the bridge: cement, wood, steel, and asphalt," stated *Germany's Highways*. (HHA, Berlin)

Left: Iron rods are placed before the cement is poured. This is what is termed "reinforced concrete" in the United States today. (Kobierowski, Berlin)

"The leveling work on a bridge has to be prepared and performed with great care, so that the deck lays perfect on the bridge… The road then continues to Zwing. The 8-km stretch, almost completed in the last few weeks, was started in the spring of 1936. About 450 workers were employed there. It was to be completed and ready for traffic in the spring of 1937. The partial stretch Bayrisch-Zell–Jatzelwurm–Niederaudorf was also completed by about 450 workers. The height difference between Bayrisch-Zell and Sudelfeld is 320m. It will be overcome with an ascent of no more than 10% by a 4.3 meter road," stated *Germany's Highways*. Note the steamroller in the background. (Press Photo Central, Berlin)

The first Reich Autobahn (RAB), from Frankfurt am Main to Darmstadt, photographed in May 1935. (Associated Press, Berlin/LC)

"The Denkendorf Valley Bridge viewed from the side, fall 1935," as seen in 1937's *Germany's Highways*. The author's account of one worker who lived in one of the prewar autobahn camps described how: "All these buildings were clustered together like a small village around a grassy area with flower beds. In such a surrounding, even those men embittered by long unemployment and skeptical of everything came to feel at home, particularly since on weekends they could return to their family and friends; and, on top of all this, they were receiving good pay." (Kobierowski, Berlin/LC)

is the Brown Memorandum of December 1932. When Hitler made his plans known in February 1933, Rudolf Hess mentioned Todt and his work. Hitler was impressed, and called for Todt.

On May 1, 1933 at the Berlin Tempelhofer Field event, orchestrated by Dr. Speer, "Hitler said before a crowd of hundreds of thousands, 'We are developing a program for the construction of new roads, a gigantic undertaking that will need millions. We will remove any opposition! It will help reduce unemployment.'"[4]

In June, Hitler promulgated the law to create a Reich Autobahn Agency (RAB) under Dr. Julius Dorpmüller, and appointed Dr. Todt as General Inspector of the German Road System, making him, in effect, "supreme commander" in all matters relating to highway planning and

CARE OF THE RAB CONSTRUCTION WORKERS

Initially workers were able to travel from their homes, but as construction of the RAB continued, it was necessary to build camps for workers. The first camps, built in 1934, caused complaints among the workers, so Todt stepped in to ensure proper accommodation was provided, making an appeal directly to Hitler. Todt wrote:

> It was decided, therefore, to make an appeal directly to Hitler. When the appeal was made, the Inspector learned that the Führer had already designed proper housing for the workers, housing which included clothes lockers, hot and cold water, bedrooms, and living rooms. The plans were soon placed into effect, and the workers had better housing than those of any other European country.[5]

Two million Reich Marks were put aside for the barracks, and Social Services, along with the DAF and "Beauty in Work," undertook their construction. Each camp housed around 216 workers as Vosselmann describes:

> Each unit consisted of 12 bedrooms with 18 workers each. Each worker had a bed, cabinet, chair, and place at the table. There was a common room, good lighting, and large windows with simple but artistic pictures on the walls, a stage, etc. The common room was also used for meetings.
>
> In the larger camps, films were shown and theatrical performances were given. Each camp had a green area, showers, kitchens, and snacks to keep up strength. Some of the camps even had dietary provisions.

Todt created an office to oversee the maintenance of the camps. Health systems were put in place, and victims of accidents went to a specialist hospital at Hohenlychen. "Strength Through Joy" helped run a cultural program at the camps, theatrical shows, films, and traveling libraries. The periodical of the General Inspectorate was of course available at all camps, along with other newspapers and educational materials. Money for board and lodging was automatically taken from the workers' pay. Free railway tickets enabled married men to visit home every fortnight, and single men once every six weeks.

As James Shand noted, "Some sense of what camp life meant ... emerges in the following remarks written almost 40 years later by a resident in the Bavarian labor camp: 'In May 1937, upon entering the work camp, I found there such living conditions as I would describe even today as absolutely model... The camp consisted of several 80-man living barracks, a similar building with showers and toilets, a canteen with a well-run kitchen, plus reading and writing rooms, as well as our own infirmary...'"[6]

Germany's Highways boasted: "For the whole building industry, the social conditions of the RAB became an example."

A typical RAB workers' barracks and flagpole with Nazi swastika banner. (CER)

Top right: A state-of-the-art RAB workers' camp kitchen, complete with stainless steel fixtures. The workers were served one hot meal, with meat, every day. (LC)

Middle right: RAB workers wash up in a local camp barracks. Normal work weeks were 48 hours, however workers received pay for a guaranteed 32 hours a week, even when bad weather prevented them working. This meant that workers had an income of at least 30 RM. (LC)

Below right: The RAB barracks' sleeping quarters held 18 workers apiece. (LC)

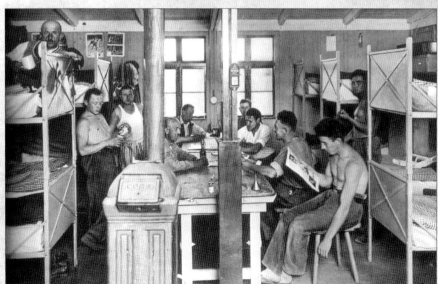

THE IRON MEN OF THE FÜHRER'S ROADS

Photographs of some of the men who worked on the autobahns, taken during the 1930s. From left: An old shaft master of the Bayreuth–Falls autobahn stretch. For 43 years he was involved in road construction in all parts of Germany, but never saw an undertaking like the autobahn (*Archiv, G.I./OT/LC*); a young worker from Heidelberg—strong, proud, and full of life with his 23 years (*OT/LC*); wages are handed out to workers. (*Archiv, G.I./OT/LC*)

construction under the new Third Reich, second only to the Führer himself. This task would occupy Todt full-time for the next five years. The Nazis took over HaFraBa, renaming it the Gesellschaft zur Vorbereitung des Autobahnhaus, or GeZuVor for short. GeZuVor had responsibility for funding the building project. Membership was open to everyone, and Göring helped bring membership to over 3,000—which assured financing.[7] Todt's organization was to become a powerful institution in the Third Reich, a real counterpart to the Deutsche Reichsbahn (German Reich Railway) as an influence on the national transport infrastructure.

In the wake of the resignation of the former HaFraBa public relations director Willy Hof, Dr. Todt in effect became State Secretary for Roads to Hitler, much as Ernst Röhm was SA Chief of Staff to the Führer as the Supreme Commander of the SA Stormtroops, a strategem again adapted from Fascist Italy.

The country was divided into 11 GeZuVor sections. Each section was responsible for investigating the condition of roads, rivers, and canals, and made their own independent plans that were then submitted to the government for approval. Plans were public, and objections could be made. Todt had the final decision on all complaints. He finalized plans, and ordered construction to start. Initially six stretches of road were approved, a total of 6,900km (4,287 miles). These routes were:

1) Lübeck–Hamburg–Hannover–Kassel–Frankfurt am Main;
2) Stettin–Berlin–Leipzig–Hof–Nuremberg–Munich;
3) Aachen–Cologne–Magdeburg–Berlin;
4) Saarbrücken–Kaiserslautern–Mainz–Frankfurt am Main–Fulda–Erfurt –Leipzig–Dresden–Breslau–Beuthen;

The pillars of some bridges were faced with stone. "That means work and bread for many needy areas," stated *Germany's Highways*. Though using the most modern technology, the autobahns were to be, as Shand put it, "bold and sweeping, but simultaneously to blend with the landscape." (Shand) (HHA)

5) Saarbrücken–Landau–Bruchsal–Ludwigshafen–Stuttgart–Ulm –Munich–Berchtesgaden; and
6) Hamburg–Wittenberg–Spandau–Berlin–Glogau–Breslau.[8]

Later that year, work began:

More than 30,000 workers were supplied for the task, and within the next several years, the labor force was increased to 70,000. Hitler was especially interested in east–west highways in order to meet the demands of a two-front war... His goal was a network of 7,300 miles of four-lane highways. A quarter of this network was completed by 1938. Today, many Germans remember Hitler as "the man who built the Autobahn."[9]

Hitler himself inaugurated the new Nazi version of the autobahn with the famed "first cut of the spade" near Frankfurt am Main on September 23, 1933, a truly ironic weekly newsreel photo opportunity for the Nazis, considering their earlier opposition to both Stufa and its successor, HaFraBa.

"This giant, 25-meter-wide machine cut the Oder River marsh, and the resulting canal became the roadbed for the autobahn. A dam made of sand was created in this enormous water trench, which safely carries the autobahn," stated *Germany's Highways*. (OT/LC)

"The Sulzbach Viaduct under construction; on slender, elegant pillars it raises the autobahn over the valley," according to *Germany's Highways*. Shand discussed the routes chosen for the new autobahns, "The RAB network laid out and begun by the Nazis after 1933 coincided in its route alignments almost exactly with the earlier plan prepared by independent technicians and planners. This similarity suggests that geographic, demographic, and economic factors—not military considerations—dictated the placement of the autobahns." (Matthaus, Cologne/LC)

GROUNDBREAKING AT FRANKFURT AM MAIN

"An idea becomes a fact, masterminded by Dr. Todt," trumpeted the 1943 Nazi propaganda film glorifying the achievements of the General Inspector of the German Road System. The German language work by Arend Vosselmann added: "The first thrust of the spade on the Reichsautobahnen conducted by Chancellor Adolf Hitler personally was at Frankfurt am Main on September 23, 1933."

According to *Danzig Report No. 106*, at least two special-issue stamps were printed showing the Führer and Reich Chancellor breaking the first spadeful of ground on September 23, 1933 on the event's third anniversary in 1936.

Bottom left: The Führer (right) removed his cap as Dr. Todt (left) briefed him from a schedule of the day's ceremonial events, as they were to unfold. Governor Sprenger stood between them at center left. This is the only known event during the period 1933–1939 where Hitler was photographed without a swastika armband on his left sleeve. This may have been a deliberate choice, to show that he was building the autobahn not as the head of the national ruling party alone, but rather as the leader of all the people. (HHA)

Above left: From right to left, Governor Sprenger, Army Commander in Chief Gen. Werner von Blomberg, Dr. Todt, Hitler, and other officials move toward the groundbreaking site. (HHA)

Below: Hitler shoveling the earth for the ceremonial groundbreaking, his entourage of top Nazi leaders looking on. This image appeared all over Nazi Germany in books of photos. Dr. Todt grins behind him. (HHA)

"Thousands of hands cleared the terrain for the autobahn. Now, the modern machines move in to pave the roadbed," stated *Germany's Highways* in 1937. (People and Reich/Berlin/LC)

Snyder wrote that the autobahns were, "An advanced system of highways constructed by Hitler as 'an overture of peace,' but obviously undertaken to permit the rapid movement of troops in war."[10] In addition to the military rationale, the autobahn building program served two other purposes of the Nazi regime: to put unemployed Germans from the Depression era back to work, and thus revitalize the German economy, and to promote automobile usage by the general population.

Neil Gregor in *Daimler-Benz in the Third Reich* stated: "The Autobahnen projects probably did more in the short term to alleviate unemployment than they did directly to stimulate automobile sales, although here, too, their importance should not be underestimated. The roadway program was nonetheless crucial."[11]

As reported by the journalist, Max Domarus, at the "first cut of the spade" in Frankfurt in 1933, Hitler made a speech which presented his theory that the program would create work and increase consumption:

Today we stand at the threshold of a tremendous task. Its significance not only for German transportation, but, in the broadest sense, for the German economy, too, will come to be appreciated in full only in the course of future decades. We are now beginning to build a new artery for traffic!

...I know that this gigantic project is only conceivable given the cooperation of many; that this project could never have evolved had the realization of its greatness and the will to turn it into reality not seized hold of so many, all the way from the Cabinet and Reich government, up to the German Reichsbank and the German Reichsbahn.

"With unheard of rhythm, the autobahn blends into the hills and forest," stated *Germany's Highways*. (Krajewski, Berlin/LC)

At the same time, we are fighting the most severe crisis and the worst misfortune which have descended upon Germany in the course of the past 15 years. The curse of unemployment—which has condemned millions of people to a degrading and impossible way of life—must be eliminated!

...What is beginning today with a celebration will mean toil and sweat for many hundreds of thousands. I know that this day of celebration will pass, and that the time will come when rain, frost, and snow will make the work trying and difficult for everyone. But it is necessary. This work must be done, and no one will help us if we do not help ourselves!

In my view, the most productive way of leading the German people back into the process of work is to once again get German industry going by means of great and monumental projects. In taking on a difficult task today ... you are ensuring that hundreds of thousands more will receive work in the factories and workshops by virtue of your increased buying power.

It is our goal to slowly increase the buying power of the masses, and thus to provide orders to the centers of production, and get German industry off the ground again...[12]

In 2007, the view that the autobahns alleviated unemployment was challenged by Prof. Adam Tooze in his groundbreaking work, *The Wages of Destruction*. He asserted that, "The autobahns were never principally conceived as work creation measures, and they did not contribute materially to the relief of unemployment... They followed a logic, not of work creation, but of national reconstruction and rearmament, a logic, indeed, that was as much symbolic as it was practical."[13]

"Will is made reality! The autobahn was ceremoniously turned over to traffic," stated *Germany's Highways*, "in several ribbon-cutting openings such as this one, as NSKK cars and motorcycles parade down a stretch of roadway." (Fels, Stuttgart/LC)

The autobahn off-ramp in Mannheim, where Dr. Speer was a member of the black-coated SS. "Only a few lanes, but large and monumental. The simplicity indicates the thoughts behind it," stated *Germany's Highways*. Today, such a sight is commonplace around the world, but in those days a structure like this was considered revolutionary. (Schmolz, Koln/LC)

Tooze explains that, "In his seminal memorandum on *Road Construction and Road Administration* of December 1932, Todt presented the program of road modernization—not as an answer to the crisis of unemployment—but as a means of national reconstruction."[14] Todt suggested that there should be built an integrated network of 6,000km [3,728 miles] of new roads over five years. He made clear that the ultimate rationale for these roads was military, as Tooze describes: "On Todt's motorways, 300,000 troops could be ferried from the eastern to the western border of the Reich in two nights of hard driving. From its inception, Todt's vision was thus intertwined with the dream of national rearmament. An army of 300,000 was three times the limit stipulated by the Treaty of Versailles."[15]

As Zentner and Beduerftig explain: "Short-term troop movements could be achieved more quickly on the autobahn than by rail. Thus,

DEPICTIONS OF THE AUTOBAHN

Noted Vosselman, "In many sections beginning with the Königsberg–Cologne route, the autobahn was depicted in both color painting and black-and-white photographs. Drawings and models of the bridges, paintings of sections and landscapes, plus interior pictures of rest houses and gas stations completed the overall impression of beauty seen in the various RAB art shows."[16]

Instigated by Todt, an exhibition was held in Munich between June and September 1934 entitled *Die Strasse (The Roads)*. Vosselmann explains: "Eight monumental pictures were exhibited with the theme *German Roads in Changing Times*. Individual titles were *Old German Bohlenweg, Roman Roads, Roads at the Time of the Nibelüngen, The Roads of the Middle Ages, Roads in the 30 Years' War, Time of Romanticism*, and the *RAB*."

In 1936 another exhibition was held in Munich called *The Roads of Adolf Hitler in Culture*, which showcased the work of around 200 artists. Most of the paintings depicted the autobahn. Besides these

"A high-speed, divided highway, the autobahn winds through the Werra Valley south of Norheim in this 1938 photograph," stated the Time-Life Books series *The Third Reich*. (LC)

A good view of Nazi Germany's prewar autobahns still under construction. "Dr. Alwin Seifert said in 1943, 'We must close the gap that was caused by years of errors in regards to nature and technology. We succeeded in completing 1,000km of motor vehicle roads, the mightiest work that technology produced in such a short time. Every driver must feel that every moment of driving on the RAB, he knows what area he is driving past, such as Lower Saxony and the Thuringian Forest, which is different from an Alpine forest, in the all-familiar German homeland.'" (HHA)

Below: A 1938 cover of *Deutschland* (*Germany*) magazine, the Nazis' English-language periodical aimed at western readers. This one touted the autobahn. Noted Vosselman: "In the north, Bremen and Hamburg are connected by about a 70-km line, while Hamburg and Lübeck will be connected soon. The Heath can now be seen as never before! The two most important German port cities—Hamburg and Bremen—are now more closely connected by the autobahn." (Courtesy George Peterson, National Capital Historic Sales, Springfield, VA)

Right: A good aerial view of the Munich–Salzburg autobahn that crossed the German-Austrian frontier, from *Germany's Highways*, 1937. The book said of the road-building in Austria: "The mostly mountainous area with its poor productiveness was not conducive for tourist traffic and industry. Lastly, because of its seclusion, its nationalism hindered its economic growth. Therefore, it was the role of the government to help this impoverished area, and to start as soon as possible the Bavarian–Austrian Road." (LC)

exhibits, the works were also put on display in various other places, and published in books that dealt with the same subjects.

In 1940, Todt asked watercolor artist Ernst Huber to paint a series of paintings of the RAB. Within a year he had completed at least 162 paintings, and an exhibition entitled *A Painter Experiences the RAB* was opened by Todt in November 1940.

The RAB was also depicted through photography, both for experts, and the general public, in many publications. The most famous of these was *Die Strasse (The Roads)*, published by Todt's department. Todt himself was a talented photographer and he made a point of only hiring the best photographers to illustrate the autobahns in magazines or books.

General Heinz Guderian could declare in 1940, 'We have enjoyed the blessings of the Reich Autobahn on the march to liberate Vienna, and then on the march to the [Czech] Sudetenland, on the march against Czechoslovakia, against Poland, and against the Western powers. What a joy to march within Reich territory!'"[17]

Shand notes that, "There were plenty of independent reasons for building the autobahnen as rapidly as possible and on the routes they traveled. Any possible resulting military benefits, though not ignored, were accidental, if welcome, by-products." Contemporary observers were not slow to fear the military utility of the autobahns: "Foreign observers frequently depicted the new German motor roads as ominous military developments... By 1939, an American military engineer, noting that Todt was now in charge of the new Rhineland fortification project [the West Wall] described the autobahns as 'arteries for the highly mechanized German automotive army.'"[18] However their unfinished state at the beginning of World War II, and the fact that some roads were completely unused during the war, did lead some later observers to doubt the military rationale behind the system.

This primary military rationale did not of course preclude peacetime use; and the autobahn program did create work. Todt estimated "that an annual budget of 1 billion Reich Marks would enable him to employ 600,000 workers, especially if the use of machinery was kept to a minimum."[19] Tooze concludes: "In practice, however, the effect of the Autobahn program on unemployment was negligible. In 1933, no more than 1,000 laborers were employed on the first autobahn section. Twelve months after Todt's appointment, the autobahn workforce numbered only 38,000, a tiny fraction of the jobs created since Hitler took office." Vosselmann has different figures, concluding that, "Only about 250,000 people maximum were employed by the RAB. Of those, 130,000 were direct at the sites, and 120,000 indirectly, delivering material, etc."[20]

Of course the RAB could not solely employ unemployed workers. In fact, "many changed jobs in order to work on the RAB. After just a few years, there was actually a shortage of workers, and machines found a new field: German industry that developed tools and machines to move earth..."[21]

Building the RAB
When Todt was appointed as General Inspector, Hitler outlined his plans for the road system to him, as Todt later recalled:

> Shortly after the General Inspector of Road Building had been named, Hitler outlined his plans for a system of highways. He invited the Inspector into his park [at the Old German Reich Chancellery in Berlin], then spoke for over 90 minutes on highway construction.
> The Führer mentioned commitments, builders, and workmen needed to complete the task, and techniques which would result in

Germany's Highways extols the virtues of the roads: "Autobahns are the fastest connections for the traffic. Roads without obstructions save about half of the driving time... The autobahns are mostly meant for heavy motor vehicle traffic, but they are not racetracks! Wherever possible, the roads were laid out to provide a pleasant view for the driver. It depends if he wants only to see quickly passing km markers or line markers, or a nice view of valleys and mountains, forests, and trees. The creators of the roads not only did not want them used for selfish reasons, but that they are used for new experiences." Some foreign visitors in the 1930s noted how empty some of the autobahns seemed. Hitler had admitted in early 1933 that the

new roads were not justified by the number of cars in the country at the time. In the secret Spandau diary entry for February 21, 1948, inmate Speer noted that Hitler once asserted: "It isn't inventory that creates demand, but demand is created by changing circumstances. Four years ago, when Dr. Todt and I laid down the route of the autobahn from Munich to Rosenheim, somebody told me that there's no traffic there. I told that petty-minded person, 'Do you think there'll be no traffic once we have the autobahn completed?'" His hope was that the "people's car" would mean many families could afford a car, however, delays in construction meant that many of the autobahns remained quiet. (Wipfler, RAR, Berlin/LC)

quality construction. In closing, he made this statement: "I have a deep understanding of the great need for highways, and I have faith that you will direct the program with honor and enthusiasm."[22]

Todt started detailed planning in July 1933, keeping Hitler closely abreast of developments:

Although the entire road plan was Hitler's—and although he took the first step in the organization of the plan—he allowed the General Inspector and his staff to make all decisions. He considered it a good policy to be in agreement with those in daily contact with the problems of highway construction.

He did, however, offer suggestions when problems were brought to him. The bridge symbolized the Führer's approach to construction. He is concerned with beauty and appearance, but the primary goal is durability. As he has so often stated, "What we build must stand long after we are no longer here."[23]

Dr. Todt's position meant that he was not just creating the autobahns, but improving the entire road system. There were basically three different classes of roads: National Roads, First Class Roads, and Second Class Roads. There were also the small country roads that linked a host of communities across the land.[24] Repair and reconstruction of existing roads was to be undertaken alongside the construction of new roads.

In spring 1934, preliminary guidelines for construction of national roads were published. Before it could be decided what work needed to

"The Mangfall Bridge of the RAB. The wide opening of the bridge allows a great view of the valley. The bridge appears to be a continuation of the autobahn that hangs in the air…," asserted *Germany's Highways*. (Kobierowski, Berlin/LC)

"The Bergen Viaduct, with a view from the southern side... Motor vehicle roads, identical to the initial network of the RABs, are proposed at a length of 1,000km. The first 1,000km were complete and ready for traffic in 1936. According to current plans, there is about 60–100km distance between one autobahn and the next. This space has to be filled by the highways, whose building will cover about 20,000km. This is more than 30 times the length of the autobahn network," said Germany's Highways. (OT/LC)

be done, both on building new roads and also repairing old roads, detailed surveys on the condition of existing roads were carried out by the GeZuVor sections.

The autobahns were the most important part of the new national road system, and thus a priority for Todt. He was responsible for the basic standards of design, working with the regional offices:

Dr. Todt based the selection and survey of routes on considerations of traffic flow, both actual and potential, economic requirements, and topographical details. He also decided the basic standards of design, but modifications were made to meet local conditions, and these were controlled from 15 provincial offices. It is significant that the scheme was operated by railway engineers rather than by civil engineers, and the autobahnen incorporated several characteristic features of railway layout, such as separated tracks and a long, uninterrupted stretch for high speed.[25]

"Seehamer Lake was hardly ever visited but is today open to drivers for the view and a refreshing dip," explained Germany's Highways. (HHA)

Workers check the connection between girders on a bridge using a modern work tool, the X-ray. (OT/LC)

Todt was responsible for carrying out Hitler's wishes that the autobahns blend in harmoniously with their surroundings and enhance the landscape. Like Hitler, Todt "conceived of the road not just as a physical object, but as a philosophical concept and—like a good Nazi— he defined the concept in racist terms."[26] As he wrote in the 1937 book *Germany's Highways*:

> Roads are cultural possessions. Every road that we use has its own century-old history and meaning. A road is a work of art. It arises from the engineer's creativity.
>
> The service of the road purely as a means of transportation is not the only purpose of German road construction. The road must be an expression of the environment, and an expression of German essence.[27]

In order to create fitting roads for the Reich, he assembled a team of landscape designers and forestry experts as well as architects; in particular, the architect who would be responsible for many of the autobahn bridges, Paul Bonatz and landscape architect Alwin Seifert. Todt's role was to bring together all the necessary people: architects, structural engineers, landscape, and forestry specialists, to ensure that each section of autobahn was thoroughly and realistically planned before work commenced. As Shand notes:

> The most striking step toward this goal was the appointment within each construction district of a counsel for the "landscape." His overall task was "to ensure ... that the autobahns blend organically into the landscape" by minimalizing the damage which the massive earth-moving projects were bound to inflict on fields, forests, and

"NSKK motorcycle riders in a picture of combined discipline and strength on the autobahn," stated *Germany's Highways*. (OT/LC)

AUTOBAHN BRIDGES

The autobahns required the building of hundreds of bridges, not only to cross rivers and valleys, but also underpasses and overpasses so that other roads did not cross the new highways. Professor Paul Bonatz, employed by Todt as the artistic advisor for the bridges for the autobahns, influenced a whole generation to build bridges that harmonized with nature.

Dr. Paul Bonatz (1877–1956) was a member of the Stuttgart School, and also a professor during the war at that city's technical university. Although not a member of the Nazi Party, Bonatz designed autobahn bridges and received railroad commissions. A vocal opponent of Dr. Paul Ludwig Troost's Königsplatz (King's Plaza or Royal Square) rebuilding in Munich, Bonatz thus fell afoul of Hitler, and was investigated by the SS for his criticism of the Führer and also for his alleged aiding of Jews. As a conservative neo-Romanesque style architect, Bonatz advocated historical, nationalistic architecture. Due to a dispute with Hitler over his plans for the Munich Railway Station, Bonatz relocated in 1940 to Turkey. He returned to Germany in 1947, working on the rebuilding of both Stüttgart and Düsseldorf.

Vosselman noted: "It was Professor Paul Bonatz ... who mostly through opinions, sketches, publications, and training, made the utobahn bridges aesthetically appealing."[28]

As Spotts explains:

> Bridges were a key element of the autobahn myth. Although Swiss and Italian bridge builders sometimes faced more challenging technical problems, their German counterparts surpassed everyone else in their aesthetic accomplishments. Paul Bonatz was the master. His bridge over the Lahn River at Limburg and another over a valley along the

Above: A good panoramic view of a prewar German autobahn stone bridge. Some of the major structures included the Lech River Bridge near Augsburg on Route A to Ulm; the Kaiserberg (Emperor's Mountain) Bridge over the Ruhr River on the Duisburg–Biefeld route; the Bergen Bridge over the Bergen Valley between Munich and Salzburg that was built between September 1933 and fall 1935; and the Sulzbach Valley Bridge on the Stuttgart–Ulm route that was built during 1937 and which cost 2,500,000 RM. (CER)

Left: This wartime 1940 painting by artist Carl Prötzen entitled *The Führer's Roads* shows an autobahn bridge under construction on the Seiden. (LC)

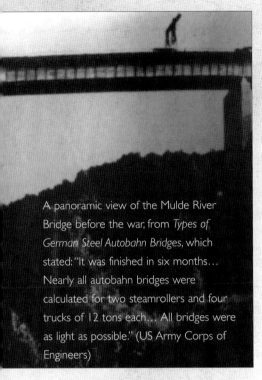

A panoramic view of the Mulde River Bridge before the war, from *Types of German Steel Autobahn Bridges*, which stated: "It was finished in six months… Nearly all autobahn bridges were calculated for two steamrollers and four trucks of 12 tons each… All bridges were as light as possible." (US Army Corps of Engineers)

"The largest spans of continuous plate girder beams are 108m and 125.28m. For greater spans, trusses have been designed in order to avoid excessive depth and thickness of the web plates. A particular solution is the reinforcement of plate girders by light arch beams," as seen here. "This kind of construction permits the covering of spans of more than 100m. Deck trusses are preferred for truss bridges, since the upper flanges of the plate girders form the railing of the roadway," noted the 1947 postwar report *Types of German Steel Autobahn Bridges*. (US Army Corps of Engineers)

Stüttgart–Ulm autobahn are examples of how he harmoniously balanced monumentalism, care for the environment, and technical skill to produce a stunning architectural work.[29]

Vosselmann observed: "The taut steel beams and the steel-concrete pillars characterized the flow of the autobahn very clearly. The Drachensteiner Viaduct on the Stuttgart–Ulm autobahn stretch—whose architect was Paul Bonatz—is distinguished by its size and boldness."[30] A US postwar document describes the Mangfall Bridge as "an engineering monument, which was one of the most beautiful of its kind in the world."[31]

The main materials for bridge construction remained natural rock, concrete, and steel, as natural stone was scarcer, though some bridge pillars were faced with stone.

The Führer's own sketches for the projected suspension Elbe River Bridge at Hamburg, all 1936, from *Hitler as Artist*. Dr. Speer said of these drawings: "The towers were planned to be over 100m high: the bridge and towers were to surpass the Golden Gate Bridge in San Francisco. Hitler abandoned his original concept of constructing the towers out of steel. Two versions of the original steel towers can be seen on the left." Hitler realized that the commercial Hansa League city of Hamburg was Nazi Germany's main overseas gateway, and thus—as early as 1934—proposed both a suspension bridge there over the Elbe River and a tunnel under it. He also wanted to construct the world's tallest building at Hamburg, to be Modernist in design. Hitler disapproved of the city's bridge plans unveiled in 1936, and in 1937 drew his own; a bridge that would be bigger than that at San Francisco. "Since soil conditions made this impossible in length, only through total surface could it be achieved. 'There can be no doubt,' Speer later said, 'that Hitler attached very great importance to this bridge, which for him was one of the most important structural documents that he hoped to build in his lifetime.' "It was to outdo America… as Hamburg itself was intended to surpass American standards.' Although soil tests were conducted and trial piers laid, the project never reached the construction phase. The skyscraper failed even to reach the final planning stage. Hitler disliked the design, and—unlike Manhattan—Hamburg did not rest on rock, and could not support the weight of the planned structure." (Speer Collection/BFP/LC)

streams. In particular, he (Dr. Todt) supervised the replacement of soil and the replanting of vegetation along completed stretches of highway.[32]

Furthermore, "Todt himself did not hesitate to cut off funds for a construction project in Baden which repeatedly failed to heed his environmental guidelines."[33]

There is no doubt that the Autobahn program was a massive undertaking as Zentner and Beduerftig noted:

> In terms of construction technology, most of the autobahn had a breadth of 24m (87.7 ft) at the crown, divided into two lanes, each 7.7m (24.6 ft) wide. They were separated by a middle strip 5m (16.4 ft) wide. Each side had an edge strip of two meters (6.6 ft).
>
> The roadway consisted mostly of tamped concrete 20cm (7.8 inches) thick. Autobahn routes were to blend into the landscape, harmoniously "swerving" or with "bold curves," and were conceived as gigantic total artworks.
>
> Bridges, in particular—whose architecture was oriented on models of the 19th century—were meant to be massive works of art and "symbols of eternity." Thus, most were constructed with hewn natural stone," making for an overall effect of beauty and strength.[34]

Not surprisingly, to build the roads and their associated structures required high quantities of materials, as Arend Vosselmann outlined:

> Each day, 2,250m³ of concrete was used. In the documentary film *The RAB* (1984–85) by Hartmut Bitomsky, an underground construction engineer who was responsible at that time for the procurement of material for 300m of roadway, reported how difficult it was to get the needed amounts of sand, gravel, etc., on the open market.
>
> Also built were 88 gas stations. Altogether, 60,000,000m² roads were covered, 280,000,000m³ earth and rocks were moved, and 152,000,000m³ of ground and soil were carried away.[35]

Germany's Highways stated that by 1937:

> About 4,000,000m³ of concrete was needed; 180,000 tons of steel construction, and about 180,000 tons of iron. Over 100 bridge constructions are in progress and/or completed.
>
> Although the usage of machinery was limited, there was an enormous need for about 50,000 rolling carts; 3,000 locomotives; 3,500km of tracks; 300 bulldozers, and 1,000 cement mixers and other equipment, like cranes.[36]

Although Todt did not employ as many workers in the program as he had stated he would, it is true that large numbers of workers were

Dr. Todt and his staff on a job-site, 1933, with an overpass behind them, in a previously unpublished photograph. According to *Germany's Highways*: "On August 18, 1933, the Society for the Preparation of the RAB, the GeZuVor, was founded, a conversion from HaFraBa, and was responsible for all technical, business, traffic, politics, and propaganda work, to complete the RAB... In one major speech, Dr. Todt stated, "The RAB is a symbol of the Nazi movement: hard-fought unity, and characteristic will to work of that movement under the Führer, Adolf Hitler. With this symbolic meaning, these roads call on technicians to find solutions from the beginning to the finish. The special consideration for the land and scenery is symbolical of the thoughtfulness and sense of responsibility toward the people. The assignment of the *RAB* is to become the roads of Adolf Hitler. It is the first work of technology that carries his name, to honor him, not only for today, but for generations." (HHA)

"Hitler inspecting highway construction," from *Hitler's Propaganda*, accompanied by Dr. Todt (left, center). "Thus, the problem of a lot of the unemployed was solved, at the same time the road system improved considerably under the direction of Doctor of Engineering Fritz Todt. A lot of technical difficulties had to be overcome, as, for example, during the construction of the German Alpine Road, leading from Inzell in Bavaria, up and down the mountains, to Berchtesgaden." Stated *Germany's Highways: Adolf Hitler's Roads*: "It soon was known that Dr. Todt was the strongest protector of the homeland. The German Forest Union was honored to have Dr. Todt as an honorary member... About the monumental assignment that Hitler gave him when he made Dr. Todt General Inspector for the German Road System, the Führer said: 'The commission to build a network of connecting autobahns offered tasks that architects wished for over many years. A uniform traffic network to be completed in the shortest possible time with roads of hundreds of km lengths through the colorful landscape that was ever-changing, was an assignment that brought great enthusiasm.'" (HHA)

needed, not only for construction work, but also for "delivery work of all sorts, cement work, steel work, stone quarries, wood industry, railroad track industry, etc."[37] Eventually, "There came a time when there were no more workers to be had. Workers were hired from Schwaben (Swabia), and several large cities. In Württemberg, men were hired from Saarland and East Prussia. The demand for work also helped the farmers and the consumer industry."[38]

The first RAB section—the Frankfurt–Darmstadt stretch—was opened on May 19, 1935.[39] This straight road was used for high-speed record attempts by the Grand Prix racing teams of Mercedes-Benz and Auto Union, until a fatal accident in 1938.[40]

Between 1933 and 1938, Nazi Germany completed 3,000km (1,864 miles) of its domestic autobahn network, while the newly created Organization Todt combined both German military and highway planning under the overall aegis of Luftwaffe Field Marshal Hermann Göring's domestic economic Four Year Plan (FYP).

In 1935, GeZuVor was largely taken over by the new regional Oberste Bauleitungen der Reichsautobahnen (Higher Building Office of the Reich Autobahns or OBR), with Dr. Todt overseeing all autobahn projects.

The following year, Chief of the US Bureau of Roads, Thomas Harris MacDonald traveled to Nazi Germany to see "Adolf Hitler's Highways," and met in Berlin with the Führer and Dr. Todt.

On September 14, 1936, Fritz Todt gave the following address to the assembled crowds at the Party Day of Honor Nazi Congress at Nuremberg in a talk entitled *German Road Construction*:

This is the third time that I have reported on the construction of our Führer's roads to the Party Congress. Two years ago, I announced the beginning of work on these roads throughout the Reich.

A year ago, I reported on the efforts of 250,000 German workers, who, after long years of unemployment, had found work in building Adolf Hitler's roads. Today, just three years after the work began, I announce the completion of the first 1,000km of Adolf Hitler's roads!

"Give me four years," the Führer asked the German people at the beginning of 1933. None of the great tasks that the Führer's will set in motion has been neglected. After three years, unemployment has practically disappeared. The military had been restored in the third year, and I can announce that German highway construction has also become a reality!

About 600km of the Reich Autobahn (RAB) have been finished by German workers and engineers, and opened to traffic. German motorists have welcomed the new roads. Over 14,000 vehicles use the section from Munich to Rosenheim each Sunday in the direction of Berchtesgaden; 6,000 to 10,000 use the completed sections near Frankfurt, Cologne, Leipzig, Hanover, Stettin, Breslau, etc.

The completed sections of the Reich Autobahn have already become the most heavily used roads in the world.

Left: A previously unpublished photograph from the Hermann Göring Albums of, from left to right, Dr. Todt, Göring in Luftwaffe Field Marshal's summer whites, and RAD Führer Col. Konstantin Hierl. (HGA)

The capacity of the new roads is best demonstrated by traffic statistics: 1,800 vehicles, one vehicle every two seconds, smoothly travel the most heavily trafficked sections each Sunday during a two-hour period.

While the number of vehicles from both home and abroad rises steadily, the labor of 250,000 workers is adding yet further kilometers to the system. This past summer, 10 new kilometers were completed each day.

The major accomplishments to date are: about 170,000,000m³ cubic meters of earth have been moved.

Also 5,000,000m³ of concrete have been used (filling 100,000 railroad cars.) The German workers have been supported by a fleet of 50,000 trucks; 2,300 locomotives; 3,000km of track, and 1,000 concrete machines.

Not only enthusiastic German drivers, but the entire world admires this great work of German labor! Numerous foreign guests attending our Olympics [in 1936] noted our road construction. A leading American said: "Tell Dr. Todt that he is building the best roads in the world."

In addition, 23 Swedes told a daily newspaper in their capital: "The autobahns are the best automobile highways in the world." A French newspaper commented: "France once led the world in highway construction, but it has fallen far behind Germany." Members of the Swiss Automobile Club wrote: "Once a tourist has driven on the autobahn, he is spoiled."

A Danish newspaper added: "They are the expression of a national energy that compels the greatest admiration," and the headline to an article about Adolf Hitler's roads in a British newspaper says: "England Needs Such Nazi Roads!"

Only one country and one system is unable to recognize our accomplishments: Russia. They make the lying claim that our road construction program is only on paper. That gives us the right to take a brief look at road construction in the Soviet paradise! What is its plan, its accomplishments?

Right: A good period map from the May 1937 edition of *Current History* magazine's article "Germany's New Roads: Are These Nazi Highways Designed for the Traffic of Tourists, or War Tanks?" by writer Frank C. Hanighen. It stated: "The old German roads, well paved with macadam or concrete, admirably served the tourists and leisure class motorists, but they were narrow. Only two vehicles could pass in opposite directions at a time, and they featured hairpin turns… Tourists in 1939 will find a different picture… The new roads are 90% concrete, 5% asphalt, and 4% stone blocks… Are the new RAB merely luxuries for the owners of Hispano-Suizas? The French do not think so. General Serrigny, writing in the *Review of the Deux Mondes* (*Two Worlds*), points out: 'On these routes, trucks carrying 30 men and traveling two abreast at a constant speed of 38mph and spaced 15m apart would make it possible to transport 72,000 men an hour, assuming that half of the trucks were used for material. No more slow embarkations, nor tedious stops in railway stations; not even bottlenecks are to be feared. The mechanized weapons of the Army can be shifted from the right wing to the left, from one theater of operations to another with a speed unheard of before. The speed of maneuvers can be increased tenfold without increasing in proportion the difficulties of supply.'" (Enoch Pratt Free Library, Baltimore, MD)

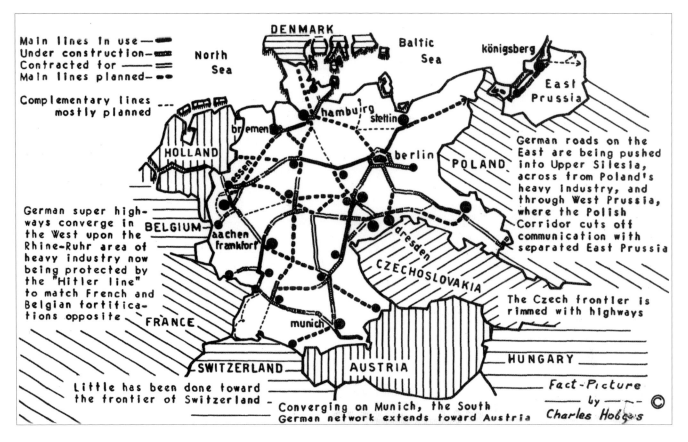

Main lines in use ─
Under construction ─
Contracted for ─
Main lines planned ─

Complementary lines
    mostly planned ---

German super high-
ways converge in
the West upon the
Rhine-Ruhr area of
heavy industry now
being protected by
the "Hitler line"
to match French and
Belgian fortifica-
tions opposite

Little has been done toward
the frontier of Switzerland

German roads on the
East are being pushed
into Upper Silesia,
across from Poland's
heavy industry, and
through West Prussia,
where the Polish
Corridor cuts off
communication with
separated East Prussia

The Czech frontier is
rimmed with highways

Converging on Munich, the South
German network extends toward Austria

Fact-Picture
by
Charles Hobgas

Control of Russia's highways rests not as before with the experts, but with the police. That is justified, since highway construction in Russia uses forced labor exclusively.

There is no free labor, or free companies. Highway labor is compulsory to 45, and each woman from 18 to 40 must give a certain amount of time each year without pay for highway construction.

More prosperous farmers and their families must give double the standard amount of labor. The newspapers at present are reporting that a major military highway is to be built connecting Moscow and Minsk; 30,000 prisoners are to do the work. Despite the massive scale of forced labor used to build Russian roads, the actual accomplishments are limited. It gives a miserable impression of the actual results of repeated five-year plans.

The enormous nation of Russia today has just 60,000km [37,280 miles] of paved roads, just 12m per square kilometer. Germany and other European nations in contrast have on average 40 times as much paved road per square kilometer as Russia! The limited amount of paved roads testifies to the low level of automotive culture in this large, yet barren and uncultured, nation.

Russia does not need more roads, since the huge country has barely 320,000 vehicles today. Russia's entire supply of motorized vehicles is no greater than the number of new vehicles in Germany over the past year and a half. The miserable situation is clear from the goals of the second five-year plan.

HERMANN GÖRING AND THE FOUR YEAR PLAN

On October 18, 1936, the Führer formally established the Nazi Economic Four Year Plan that he had earlier announced at the annual Nuremberg Party Congress.

Hitler asserted that, "Within four years, Germany must reach full independence from abroad in all raw materials that can be produced by German skill, by our chemistry, our mechanical industries as well as from our mines."

This Soviet-style drive was prompted by the severe shortage of raw materials in the overall German peacetime economy. This shortage was hindering the Third Reich's rearmament program for the European war that Hitler was planning: "The shortage of raw materials and foodstuffs—as well as bottlenecks in the supply of fuels critical to munitions production—made obligatory a decision as to the future tempo of rearmament necessary."[41] The Four Year Plan would put the German economy on a war footing, in peacetime, without a fall in the standard of living: "Hitler demanded that, '1) The German Army must be ready for action in four years, and 2) The German economy must be ready for war in four years.' Fulfillment of these aggressive and expansionist goals demanded relentless mobilization of the economy."[42]

In charge of this new, sweeping program was to be his main Nazi paladin, Prussian Prime Minister, Aviation Minister, and chief of the new Luftwaffe, Colonel General Hermann Göring (1893–1946). Under Göring's leadership, the Four Year Plan would concentrate on obtaining and allocating raw materials; monitoring labor deployment; and putting production under strict control, to ensure that priority was given to products necessary for rearmament. Hitler had already secretly recognized that this would only be a temporary solution, and that a permanent solution would only be possible through the expansion of living space, or the base of resources. Further development of the autobahn system was part of the plan, as were public works projects, both under Fritz Todt. As Zentner and Beduerftig explain:

The uncoordinated system of special deputies allowed the plan to become an aggregate of many separate measures and partial planning, so that until the war, the plan's investments lagged by 40% of its goals. Despite failures in fulfilling these goals, the plan did achieve a substantial alteration of the economic structure in favor of the production goods industries, for which the investment volume in 1939 was 250% higher than in the boom year of 1928.[43]

Opposite: Hermann Göring (giving a casual, back-handed Nazi salute, not a wave) made a major speech on the new Four Year Plan in October 1936 at the Berlin Sports Palace, with the text seen here in his left rear pocket, and Reich Foreign Minister Baron Konstantin von Neurath seated at left. It was in this speech that Göring challenged his Nazi listeners to choose either "Guns or butter!" Chortling at his own girth, Göring laughed, "Butter just makes us fat!" His listeners laughed along with him, and Nazi Germany plunged further down the path toward war. (Previously unpublished photo from the Hermann Göring Albums, LC, Washington, DC)

Göring (center in summer white Luftwaffe uniform) and Reichsführer-SS Heinrich Himmler (right) are seen in a previously unpublished view examining ball-bearings (left) and other industrial implements before the war. Later on, Dr. Speer would detail how Allied air raids on the major ball-bearing plants at Schweinfurt helped cost Nazi Germany the arms manufacturing war. (HGA)

This map illustrates the network of 1,000km of autobahns that were built throughout the Greater German Reich before the start of World War II on September 3, 1939, as seen here. In the upper right-hand corner is the Danzig section, over which Hitler invaded Poland on September 1st. According to the Hitler Historical Museum, however, "Over 2,000km were built by 1938, and today approximately 11,000km cover Germany. True to stereotypes about German engineering and maintenance, road designs are solid without a single pothole, with 4% or lesser grades, long acceleration and deceleration lanes, gentle curves, and freeze-resistant surfaces. Just as Hitler insisted on having buildings and other infrastructure that would last 1,000 years, the engineers of the autobahn design things right the first time, and perform critical inspection and thorough maintenance to keep the best highway system in the world at peak operational performance." (LC)

The amount of paved highways is to increase from 60,000 to 80,000km, that is, from 12 to 16 meters per square kilometer. The second five-year plan also proposes to establish 55 large and 172 smaller repair shops and 1,000 filling stations in the huge country. We have nothing with which to compare to the wretchedness of this program and its goals!

We do not live in the Soviet paradise, but rather in National Socialist Germany, in the Reich of Adolf Hitler. We work with free workers to carry out the Führer's tasks.

Luftwaffe Commander-in-Chief Field Marshal Hermann Göring (left) holds his informal marshal's baton aloft in his right hand as he examines a map of the RAB with an NSKK aide (left of center, wearing a gorget on his chest) in this previously unpublished photograph from the Hermann Göring Albums. Göring had four batons in total, two of which still remain unaccounted for. (Library of Congress, Washington, DC)

The Führer and Luftwaffe Colonel General Hermann Göring (center) visiting the annual International Automobile and Motorcycle Show in Berlin on February 18, 1938, where he delivered the regime's automotive policies in his opening-day ceremonial speeches. Here they view a chassis with drive train components. Development of the RAB was paralleled by the drive to motorize Germany. A variety of steps were taken to encourage Germans to take to the roads, including reduced tax on new vehicles, and cheaper cars. Todt's biographer stated that the number of motor cars rose from 41,000 in 1932 to 200,000 in 1936. (Previously unpublished photo from the Hermann Göring Albums, LC, Washington, DC)

On September 23—less than 900 days from the beginning efforts—the first 1,000km of Adolf Hitler's roads will be finished. They are roads unequaled anywhere else in the world in their technical excellence and beauty. Is this a work of technology? No! Like so much else, it is the work of Adolf Hitler!"

The 1937 book, *Germany's Highways*, placed Hitler's achievements at the pinnacle of a history of road-building by historic empires such as the Persians and Romans, and praised the progress of the autobahns:

Germany not only caught up with the foreign countries, but surpassed them with the construction of the RABs. Three years after starting, 1,000km were made available for usage, and with giant steps, they were nearing 2,000km.

When flying over the country, one can see everywhere the light-colored bands of the RAB; they lend a new face to Germany. Their imposing width, tight yet flowing turns, forces the eye in their direction. The origin and destination comes clearer into the consciousness, and we suddenly feel a greater connection.

A drive on these roads brings one to a powerful adventure: the oneness of the Reich. They follow age-old traffic patterns and directions, which for hundreds of years connected north and south to exchange goods, where Orient and Occident came together.

Also in the east–west direction, the RABs follow tradition-rich connections. These are the very paths that the political and cultural conquests of the German East followed, the paths that connected Eastern to Western Europe.

The RABs are the pioneers of new settlements. They bring new blood to underdeveloped areas and borders, and give greater security and better opportunity to keep them safe. Borders always have to be guarded, and it is a new responsibility of new traffic and area planning to create personal and economically safe positions.

Even the diagonal connections of the RABs—the section Hamburg–Berlin–Breslau, or the Ruhr area–Nuremberg–Passau—are age-old arterial roads. The Nibelüngen road was used by the fighters of the heroic saga along the Danube. On these same paths traveled the Imperial power of the Holy Roman Empire of the German nation from Aachen over Mainz and Frankfurt toward the southeast.

These two main arteries of the RAB make up the axis of the connection between the Occident and the Orient. They make up the connection between the North Sea and the Black Sea, and extend into India and the Far East. No other people in history have always found the way to India.

Alexander the Great used it, and numerous cultural monuments of the Northern Race were found on this path as witnesses to it... "Via Vita," said the ancient Romans: "the road is life." The roads of Adolf Hitler are thought out and built for the centuries. They usher in a new traffic epoch, the time of the motor vehicle. They will serve generations...

It is a fact: the whole meaning of this truly gigantic undertaking is not yet fully known, but only imagined. Only when you look at this historic period can you see the technical development before your eyes! Only to those who can think in centuries will the generality of the plan be noticeable[44]

The autobahns "developed into one of Hitler's greatest propaganda windfalls. German media had a heyday, fêting the road system as 'the greatest single masterpiece of all times and places,' 'the eighth wonder of the world,' 'greater than the Great Wall of China,' 'more impressive than the pyramids,' 'more imposing than the Acropolis,' 'more splendid than the cathedrals of earlier times.' Needless to say, it was also trumpeted as evidence of the superiority of the Nazi system over democratic government. During the 1936 Olympic Games the autobahns withstood the heaviest traffic it has seen, thousands of cars and hundreds of buses."[45]

Architects were commissioned to prepare drawings for a gas station suitable for the RAB. The first gas station to be completed on the autobahn at Frankfurt, in May 1936, was designed by Carl August Bembe. It was large by contemporary standards, and heavily used. After 1937, gas stations evolved from flat-roofed designs to several roof designs, slanted, or with wings. The gas stations were to be placed every 30–40km. The station attendant was a specially trained representative of the RAB. The photo shows how the Darmstadt autobahn gas station appeared to incoming motorists. (LC)

Having broken through the ribbon to denote the Frankfurt am Main–Darmstadt section of the autobahn open, Hitler saluted onlookers from his car, with SS chauffeur Julius Schreck at the wheel. Sitting directly behind him on the mid-car jump seat was SA Gen. Wilhelm Brückner, and behind him in the rear passenger section sat Dr. Todt. (Previously unpublished photo from the Heinrich Hoffmann Albums, US National Archives, College Park, MD)

Wartime Autobahns

Three years later, on September 1, 1939, Germany invaded Poland, and World War II began two days later with the British and French declarations of war on Germany.

During the period 1938–41, only 800km (497 miles) of highways were added to the German domestic autobahn system, as the Third Reich was simply not able to simultaneously fund both the war and the construction of the originally projected 6,000-km autobahn system. The last groundbreaking was three weeks before the outbreak of war. "After September 1939, 90% of all German vehicles were requisitioned by the government, and the RAB speed limit was set at 100mph."[46]

Construction diminished in 1940 as both workers and resources were unavailable, being instead devoted to the war. On December 3, 1941, Dr. Todt stopped all construction, as all building within the confines of the Third Reich proper ceased.

There is debate about whether, during the war, the autobahns were used for military purposes, in particular troop movements: Vosselmann states: "The autobahns were seldom used for troop movements... Military transport was the domain of the German trains."[47] American

The conclusion of the ceremony marking the opening of the Breslau–Kriebau stretch of the autobahn. In the front row from left to right: unknown, Dr. Hans Heinrich Lammers, Hitler, Dr. Todt, Hitler's adjutant SA Gen. Wilhelm Brückner, and Max Amann. Vosselmann commented that, "Intentions became reality. The year 1936 was a milestone in the development of the German traffic system. The building of the RABs—the preparation of which required much work—was the first stage." (Previously unpublished photo from the Heinrich Hoffman Albums, US National Archives, College Park, MD)

historian James D. Schand stresses that Todt felt he was building the autobahns for peace.[48] The opposite view is however expounded by Karl Larmer.[49] The author of *Autobahns of Germany* concluded:

> During World War II, the central reservations of some autobahns were paved to allow their conversion into auxiliary airports. Aircraft were either stashed in numerous tunnels, or camouflaged in nearby woods. However, for the most part, the autobahns were not militarily significant.[50]

By the time the German Army was halted by the Russian Army outside Moscow, the resources available to Nazi Germany by its only partial wartime mobilization were stretched to the limit, even taking into account the usage of vast numbers of Russian prisoners of war.

By that point, only 3,860km of domestic German autobahns were available, but much of it was war damaged, incomplete, or unusable. Hitler ordered that the autobahn bridges, already the target of many bombing raids, were to be blown up in 1945 to stop the advance of the Allies. His order was mainly ignored, and many bridges were left intact. The roads were left as they were, and proved an important part of infrastructure as Germany struggled to recover after the war. Today, there are about 6,830km of autobahns in Germany. In his eulogy for Todt in 1942, Hitler mentioned his work on the autobahns:

> "The German Reich's auto roads are—in the planning of their layout and execution—the work of this quite unique technical and also artistic talent. We can no longer think of the German Reich without these roads! In the future, also, they will find their continuation as natural great communication lines in the whole European transportation region. But what has in addition been done in Germany ... in the broadening and improvement of roads, in the elimination of bad curves, in the construction of bridges is so incomprehensible in its scope, that only an exhaustive study will permit a comprehensive and just appreciation of the accomplishment in its entirety.

His raincoat snagged on the shovel, the Führer prepares to break ground at the Walserberg in annexed Austria on April 7, 1938 for the new autobahn stretch Munich–Salzburg–Vienna. According to German author Arend Vosselmann, "In the median strip of the autobahn where the first ground was broken for the RAB in Austria, a large sculpture was to be erected. The sketch was drawn by sculptor Josef Thorak and showed four naked Titans, who moved a block of stone. The work wasn't completed." (HHA)

Highways to the Elbe River: A famous view of German POWs tramping westward to the rear, as US Army tanks and other military vehicles roll ever onward toward the east on March 29, 1945. (US Army Signal Corps)

THE AUTOBAHN AS FORERUNNER OF THE US EISENHOWER INTERSTATE SYSTEM

As noted by Tom Lewis:

> In 1936, the Chief of the US Bureau of Public Roads, Thomas Harris MacDonald, spent almost half the year traveling to conferences and road inspection tours, including one trip to Germany to see the Reichsautobahn, the impressive new roads designed and built by the renowned German engineer, the Inspector General for German Highways, Dr. Fritz Todt...
>
> Though he was impressed by the construction methods, grade separations, and superb engineering, MacDonald saw little justification for Hitler's 2,000 miles of Reichsautobahn that Dr. Todt and his laborers were creating.[51]

One person who *was* impressed by the Reichsautobahn nine years later, however, was a future General of the US Army and President of the United States, Dwight David Eisenhower. His views on roads had already been shaped by a disastrous army convoy across the United States he had taken part in in 1919. According to author Dan McNichol:

> The incomplete highway meant that 'Hitler's Road' was filled with sudden dead ends and long, partially built, disconnected sections. Even in its incompleteness, it proved to be a powerful weapon...
>
> Hitler fought hard to keep Ike's armies off the Autobahn; but once on it, the Allies literally used the Germans' own superhighways to chase them down. Once the Allies controlled the superhighways, they were able to force an unconditional surrender in just six weeks...
>
> Ike's men were astounded by the Autobahn's features. Some of them had never seen anything like the super high overpasses designed to allow German military vehicles to pass underneath. They marveled over the strength and size of the bridges, the four wide lanes, two in each direction, and the 15ft of attractive plantings that divided them. Everyone from the five-star general down to the enlisted man was impressed by the speeds at which they were able to move over Hitler's Road."[52]

Eisenhower himself remembered:

> During World War II, I had seen the superlative system of German Autobahnen—national highways crossing that country and offering the possibility, often lacking in the US, to drive with speed and safety at the same time. I recognized then that the US was behind in highway construction. In the middle 1950s, I did not want us to fall still further behind.[53]

When Eisenhower became the 34th President of the United States, only 53% of the nation's three million miles of roads were paved.[54] He introduced the idea of a new highway system in 1954. He saw a new system as a defense measure necessary in the Cold War: if nuclear bombs were dropped on the country, 70 million urban residents would have to be evacuated, an impossiblility without good road systems. The Federal Aid Highway Act of 1956 changed modern America's landscape forever: it launched an unparalleled 75,437-km (46,876-mile) superhighway system. Fully opened in 1991, the highway network celebrated its fiftieth birthday on June 29, 2006. Although the German autobahn remains the first national superhighway system ever built, the later US Interstate System takes pride of place as "the largest engineering project the world has ever known."[55]

CHAPTER THREE

THE WEST WALL: HANGING OUT THE WASHING ON THE SIEGFRIED LINE

"With shovel and rifle, all along the front"
Dr. Todt: Man, Mission, and Achievement, 1943 Nazi propaganda film

A good cutaway diagram showing a typical West Wall/Siegfried Line bunker with its above-ground gun turret cupola and below-ground support network of various systems. These included kitchen, communications, railway, recreation room, showers, infirmary, engine room, sleeping quarters, sunlamp room, and storage area. According to Gerhard L. Weinberg, it would have taken the Reich about a month to redeploy its forces from conquered Poland to the Western Front while the West Wall held out. Even before it was built, the Allies vastly over-estimated the effectiveness of the West Wall—perhaps because of what they knew of their own Maginot Line. *(Signal Magazine)*

On April 28, 1939, Reich Chancellor Hitler ascended the speaker's rostrum of the German Reichstag in Berlin to claim that "no power on Earth" could pierce the Third Reich's new West Wall defenses, constructed by Nazism's master builder and chief engineer, Dr. Fritz Todt. Recalling the Czech–German Sudetenland Crisis of 1938, Hitler warmed to his theme, continuing,

> At the time, I directed and gave orders for the expansion of our fortifications in the west. By September 25, 1938, they were already in such a condition as to surpass the power of resistance of the former Siegfried Line by 30–40 times. Since then, they have essentially been completed.
>
> At present, the sections that I later ordered to be added, running from Saarbrücken to Aix-la-Chappelle, are under construction. To a high degree, they are ready to assume their defensive role.
>
> The state in which the mightiest fortification of all time finds itself today affords the German nation the reassuring knowledge that no power on Earth shall ever be able to pierce this front...

This had hardly been the case when, in the spring of 1938, the Führer first turned his attention to the fact that Nazi Germany needed a counterweight in the west to the strong French Army and France's defensive system, the famed Maginot Line. This became particularly important when Hitler was planning the invasion of the Sudetenland. Fortifications on the western borders had to be in a fit state to allow relatively weak forces to defend them if the French attacked while the majority of the Wehrmacht was deployed elsewhere.[1]

THE OSTWALL (EAST WALL) 1935-45

The East Wall predated the West Wall. Despite the Treaty of Versailles, the Festung Pioneer Korps, (German fortress engineers), continued to maintain and improve their few remaining fortifications. Improvements were minor as military authorities were not interested in investing in fortifications, and the engineers mainly planned for future projects. When Hitler took office as Reich Chancellor in 1933, planning became action, and new fortifications were started.

Plans had been in place for strongholds on the eastern border with Poland, along the provinces of Pomerania and Silesia, since 1927, but work proper only started in 1935.

Initially, Gen. Otto-Wilhelm Foerster, the inspector of the engineer corps, was given 15 years to complete the fortifications. The East Wall included large artillery blocs, bunkers armed with armored turrets and guns, observation, mortar, flamethrower, and machine gun positions, and antitank obstacles.[2] Underground shelters, barracks, stores, and power plants were built, linked to the fortifications by underground galleries and tunnels.

In 1937, Hitler became infuriated by poor results seen on the East Wall given the money and efforts put into it. Work slowed the following year, as resources had to be shared with the West Wall fortifications. The East Wall was never completed, but it provided the Festung Pioneer Korps with valuable experience, and many of the elements served as prototypes for the elements in the West Wall and Atlantic Wall.

In summer 1944, the East Wall was reactivated, and the defenses hastily repaired, but they did not stop the Soviet armies for long.

In 1935, Hitler had fixed a period of 15 years for the completion of a general system of fortifications for Germany, based on the fortified area concept. However, "A few months later, the Rhineland was remilitarized, and the focus of fortification activity was shifted from the Polish to the French frontier, while the time allowance for completion of the plans was reduced first to 10 and then to four years."[3] Work on the defenses was very slow over the next two years, as "Construction on the Rhine itself was forbidden, and the importance of camouflage was stressed. Allocations of materials and funds for these purposes during 1936–37 were small, partly because of expenditures to construct the famous autobahns. By the time of Anschluss [the annexation of Austria], only a few hundred small concrete emplacements had been built."[4] In March 1938, Hitler

Right: A section of dragon's teeth. According to the 1971 work *The 12-Year Reich: A Social History of Nazi Germany, 1933–45*, "In June 1938, the regime promulgated a duty service law under which all workers could be 'industrially conscripted,' i.e. transferred to fortification work on the *Western Rampart* ... 400,000 workers from all over Germany had been literally dragged out of their beds in June 1938 to build the *West Wall*... The bulk of the *West Wall* labor force were men attracted by relatively high pay and opportunities for additional overtime earnings... The DAF ... entertained the huge migrant work population lavishly by putting on as many as 300 entertainments, such as film shows and concert parties, per night." (CER)

Above and above right: Constructing the antitank "dragon's teeth," and the final result. The teeth were concrete reinforced with iron rods. The channels dug between the teeth provided foundations and linked them together for added strength. In July 1939, an officially sanctioned publication announced that half a million men had worked on the West Wall; that a third of all cement mixing machines in the Reich had been used on the West Wall, and had thus far produced six million tons of cement; and that three million tons of barbed wire had been used. (CER)

approved extension of similar fortifications on the Belgian and Dutch frontiers, on a two-year program.

However, in May 1938, Hitler ordered an acceleration of the program on the western border, to create 1,800 gun emplacements and 10,000 pillboxes in four months. He then called for a linear defense system along the French and Belgian frontiers, a U-turn on his 1935 decision. This phase of the construction of the West Wall is known as the Limes Program, and it required a huge increase in the resources and manpower employed on the defenses:

> To expedite these ambitious designs, the Army was to assign more fortification engineer troops to the task. The State Labor Service [RAD] was to commit 50,000 men, and the General Inspector of Roads, Dr. Fritz Todt, was to be put in charge of "a large part of the concrete construction."[5]

These orders were passed on to Gen. Wilhelm Adam, heading Group Command 2 in western Germany, who was responsible for their execution. Adam, who favoured fortified area defenses, felt the order was absurd, and said so. Adam also had little time for Col. Konstantin Hierl, the leader of the RAD. Hitler aggravated the issue by sending Göring to inspect the western defenses. Despite being no expert on fortifications, Göring reported back that the defenses were a joke, and that it was all the fault of the Army. In reality of course, Hitler had been involved in the fortifications program throughout, had held it up with his changes of decision, and had given it little priority, until his decision to invade the Sudetenland made it an issue. However, Göring's report gave Hitler the excuse to blame everything on the Army, and shift his reliance to Todt and Hierl. This may even have been his agenda in sending Göring westward.

After more meetings and disagreements, Hitler issued a memorandum on "the question of our fortifications" at the end of June. It comprised his "principles" of fortifications, and an attack on large fortifications, ending with his order for 10,000 concrete bunkers, to be completed within the time. Hitler was determined to attack the Sudetenland that year, and needed as much protection in the west as he could get. He may have hoped that a show of activity might stop the French attacking at all.

Adam was not replaced as commander, as might have been expected. He continued to answer to OKH (High Command of the Army), while Todt and Hierl reported to Hitler. The memorandum had not set out the detail of the fortifications, so it became a dual operation, the Wehrmacht (Armed Forces) engineers working to the Army's policy, Todt building giant emplacements. Contention over the fortifications continued between Hitler and his generals.

The Führer and his entourage view a sand table model of the West Wall/Siegfried Line before the war. By September 1938, Dr. Todt had completed over 6,000 bunkers using over a million cubic meters of concrete. Seen here from left to right are unknown German Army officer, Army Commander in Chief Gen. Walther von Brauchitsch, Führer military liaison officer Capt. Gerhard Engel, Hitler, OKW chief Gen. Wilhelm Keitel, Luftwaffe Field Marshal Hermann Göring, two unknown aides behind him and Luftwaffe Gen. Karl-Heinrich Bodenschatz, Göring's personal liaison officer to the Führer Headuarters (FHQ). (HHA)

BUNKERS

An underground structure is a shelter, while bunkers and pillboxes are above-ground structures. However, since 1945, Hitler's underground air raid shelter has persistently been termed a bunker. The thousands of bunkers along the West Wall were of two main types: passive bunkers such as troop shelters and ammunition stores, and armed bunkers. These ranged from large, well-protected and expensive bunkers for positions of tactical importance, to small, lighter pillboxes.[6] They were placed to provide cover for each other, intended to work as part of a system, rather than as an isolated point of defense.

They were designed by the Army's Engineering Corps, with input from Hitler who provided sketches for blockhouses, and constructed by the OT. Before construction, a survey was made of the soil so that suitable methods were employed, and all the necessary services for the constructors had to be provided: offices, workshops, barracks, a supply of soft water, sand, gravel, and camouflage for the site.

A concrete floor was poured following excavation, then forms were erected, providing the shape of the internal rooms. Ceilings were armored with steel plates on metal beams, and a metal framework built up of a web of round steel bars (around 10 layers in 1.25m of concrete). At the same time ventilation, chimneys, cables, telephone casings, etc., were installed. The walls were painstakingly laid out with heavy wooden forms; it was crucial to get them right because of the enormous pressure of the concrete poured into them.

Concrete mixers were now placed on scaffolding above the forms, and concrete was poured in. After the concrete was dry, the forms were removed. The bunker was covered over with dirt or sand, prior to camouflage, and the internal fixtures and fittings placed.

A small casemate might take a few days to construct, while a large, complex bunker, such as a command post, might take weeks or months.[7]

Tooze summarized the situation thus: "In May 1938, Hitler had wrested control of the West Wall from the Army's engineering department and handed it to Fritz Todt, the man idolized as the master builder of the Autobahns. Todt's mission was to complete the fortifications before the outbreak of hostilities, and he was to do so regardless of cost. Göring's decree on labor conscription provided Todt with all the necessary legal powers to secure the quarter of a million workers he needed..."[8]

In July, the OKH advised OKW (High Command of the Armed Forces) that the workers had to be out of the fortified zone by September 24, in case of a French attack as the invasion of the Sudetenland took place. In August, Todt rang Adam's chief of staff to say that Hitler had refused to allow an early stop to the work. Upon questioning this, more formal orders came through to Adam that work must continue. Disagreements continued when Hitler visited the defenses in late August. Toward the end of 1938, "plans were drawn up for 14,600 additional bunkers. In 1939, the numbers were increased and more workers were involved."[9]

In his diary entry for August 14, 1938 at the Old Reich Chancellery, Capt. Gerhard Engel noted: "Uproar about the West Wall. Führer has looked over the plans and talks about delays by Army agencies. He had asked the field marshal [Göring] to look into the rights of the matter.

Sharply criticized Engineer-General Foerster. Reichsleiter Bormann was standing near the Führer. He clapped Bormann on the shoulder and said, 'If I could do as I wanted, I know to whom I would transfer the work in place of the generals, Bormann. At least then I could rely on it.' Führer did not see me, as I was standing off to one side."[10] Already, Hitler was thinking of naming Dr. Todt.

Lepage explains how the various agencies actually worked on the West Wall:

> Todt directed the 22 senior building executive committees, which were commissioned to undertake the construction work from the Central Office for Western Fortifications in the Hotel Kaiserhof [Imperial] in Wiesbaden. Along with the Army Pioneers (Engineers) and the DAF, about 1,000 firms with their depot and staff worked under his overall direction on the West Wall. A building company was created, called the Organization Todt. Todt's building company existed already, but the name Organization Todt was used by Hitler for the first time at the Nazi Party Congress in Nuremberg in 1938.[10]

Zentner and Beduerftig summarized the work undertaken between May 1938 and September 1939 on the 400-mile defenses:

> ...Around 14,000 bunkers, battle positions, and dugouts—as well as the characteristic antitank "dragon's teeth," were put into place. The

The Führer tours the West Wall fortifications in 1938 with his entourage, from left to right: Reichsleiter (National Leader) Martin Bormann, Reich Press Chief Dr. Otto Dietrich, Hitler, Army Commander in Chief Gen. Walther von Brauchitsch, Führer physician Dr. Theodor Morell, and a phalanx of unknown German Army officers. The striding officer at center (eighth from left) may be German Army Gen. Wilhelm Adam, while the SS officer with the white facings on his overcoat and wearing dress sword is Reichsführer-SS Heinrich Himmler. Hitler made four visits to the West Wall while it was under construction. (Walter Frentz)

A further view of the Führer's 1938 tour of the West Wall, showing from left to right, a pair of unknown officers, German Army Gen. Erwin von Wizleben; four unidentified aides, Gen. Keitel, two unknown men, SS surgeon Dr. Karl Brandt, Reichsführer-SS Himmler, three unidentified men, Reichsleiter Martin Bormann, unknown Army officer, Führer Reichssicherheitsdienst (Reich Security Service or RSD) man, Lt. Col. Bruno Gesche, unknown man, Dr. Fritz Todt, the Führer, and a final unidentified officer. Dr. Todt's 1943 Nazi film biography stated: "Until September 1, 1939, construction grew on the West Wall to 13,700 bunkers, with 6.5 million cubic meters of concrete. This performance was due to Dr. Todt's powerful leadership and the German construction industry, and secured the western border and also gave a free hand against Poland" for the German Armed Forces—Hitler's goal all along. (Walter Frentz)

expenditure came to around 3.5 billion RM. Materials consumed included 8 million metric tons of cement (20% of Germany's yearly production), 20.5 million metric tons of filler materials, and 9.5 million cubic meters of wood (8% of the yearly wood consumption). Every day some 8,000 railway freight cars reached the construction sites with materials (for a total of 1.01 million cars). By ship and truck, 4.5 million metric tons of material were delivered. Along with the RAD and transport organizations, some 100,000 workers from the fortification engineering corps of the Army and 350,000 from the TO were utilized.[12]

In May 1939, "Hitler believed that the time had come to renew his psychological assault on the west, in a type of war of attrition, with a tour of the 'West Wall.' He aimed to intimidate the Western powers by thus drawing media attention to the supposedly invincible, 'mightiest fortress of all time.'"[13] Hitler visited several sections of the defenses where work was completed, or ongoing, and examined the Air Defense Zone West. At the end of his tour he made a speech at Baden, to assembled generals, "expressing once more his profound gratitude to Gen. Erwin von Witzleben [commander in chief of 2nd Army Group] and the General Building Inspector, Todt... The Führer underlined once more how profoundly impressed he had been with the exemplary comportment and spirit of each member of the border guards and each West Wall worker. They had reassured him 100% in his existing belief in the present invincibility of the German West Wall."[14] Following the tour, he issued an order of the day to the Wehrmacht, which echoed these sentiments.

Left: The Führer and his entourage during the 1938 tour of the West Wall. Dr. Todt's 1943 Nazi film biography asserted that: "Todt's influence at the West Wall caused a fundamental change in the military valuation of technology. He developed a totally different style of construction at the West Wall that was geared towards military purposes… A Swedish magazine wrote in an article about the 'artist with a thousand abilities,' and called him 'the greatest employer of all times, from the first minute of the war one of the most important members of the war machine. The spade had its place of honor next to the rifle.'" Dr. Todt is fourth in line behind Hitler. (HHA)

Despite Hitler's high-profile blustering about the West Wall, in 1938–39 the French were under no illusions about German defenses in the west: in 1938, there were reports submitted by the French air attaché in Berlin that stated that the defenses were incomplete and not very strong, and that Hitler had vastly exaggerated their strength. General Maurice Gamelin, chief of the French military, accordingly planned to advance straight into Germany, rather than wait behind the Maginot Line, exactly what Hitler feared most.[15] But he didn't—not then, and not in 1939–40 either—because his resolve presupposed that the British would wholeheartedly support him, and they did not. Thus, Hitler got away with his bold—even rash—gamble over the mirage of the phantom West Wall defenses.

Many Army officers and generals did not share Hitler's faith in the West Wall. A later German Army field marshal—Erich von Manstein—wrote to Col. Gen. Ludwig Beck—Chief of the Great German General Staff—about the "overhasty" construction of the West Wall. Indeed, most of the Army generals shared their joint misgivings about war with the Allies that year; that western Germany would be both invaded and overrun by them as the Nazi armored columns smashed into Czechoslovakia.

By the summer of 1939, Todt had completed his mission. The most vulnerable sections of Germany's western frontier were reinforced with thousands of bunkers and gun emplacements:

Right: The West Wall/Siegfried Line. Though the level of defense varied according to the guiding principles of those who had constructed a particular section, on the whole the West Wall was a position of relatively light defense with great depth. It had three "lines": a forward position of 2–10km studded with minefields and antitank obstacles; the main line, between 3 and 8km deep, which consisted of the bulk of the defensive means; and finally, the rear position, a few kilometers behind the main line. (From Hitler's Siegfried Line)

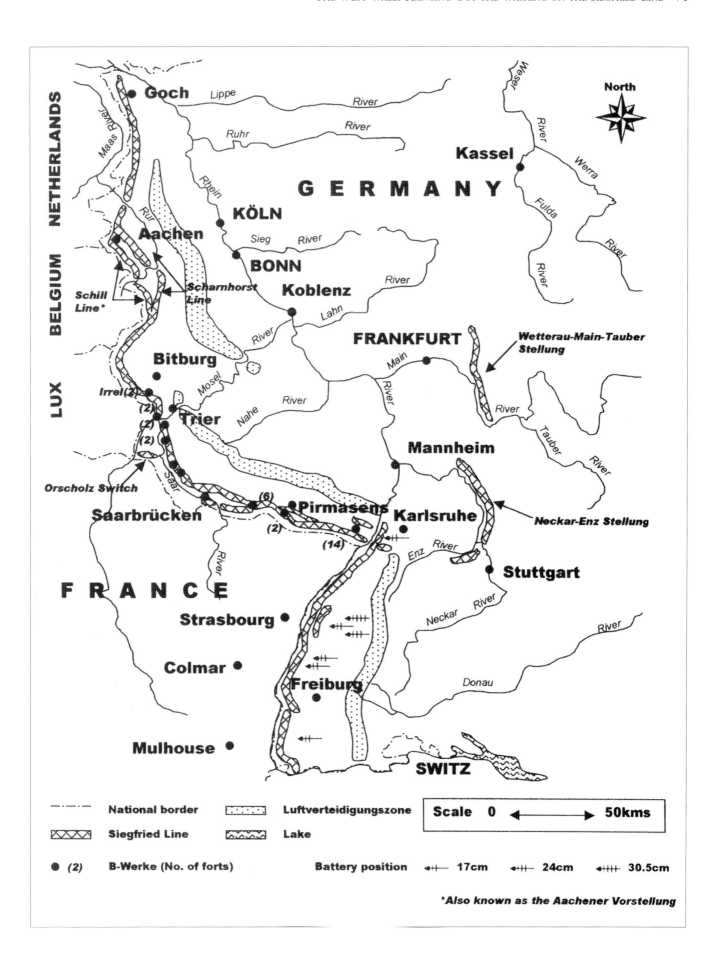

Scale 0 ◄——————► 50kms

– – – –	National border	⫶⫶⫶⫶ Luftverteidigungszone
⬡⬡⬡⬡	Siegfried Line	≈≈≈≈ Lake
● (2)	B-Werke (No. of forts)	Battery position ⊢┼ 17cm ⊢╫ 24cm ⊢╫┼ 30.5cm

*Also known as the Aachener Vorstellung

THE ORGANIZATION TODT

The Nazis were fond of naming important organizations after their top leaders, such as Leibstandarte SS Adolf Hitler, Regiment General Göring, Kampfgruupe (Battle Group) Peiper, and, in this case, the Organization Todt (OT).

The Organization Todt was set up in 1938 to construct military installations, and special roads. Workers included engineers, bricklayers, carpenters, and industrial machinists. It was administered by Todt until his death, after which it was run by Speer.[16] Zentner and Beduerftig noted: "One of the principal reasons for its creation was the need to make use of the numerous service obligations of workers and employees and the commissioning of private construction firms, according to the ordinance for 'Securing the Required Forces for Tasks of Special National Political Responsibility.'"

Because Todt was General Inspector of German Roads, General Plenipotentiary for the Regulation of the Construction Industry, and Reich Minister, his organization had a very strong position. They were neither soldiers, nor members of a Party organization, and so the OT was "largely free of bureaucratic encumbrances. Such an advantage made the OT—above all because of its great efficiency in carrying out its construction assignments—one of the most specialized organizations in the Third Reich."[17]

The first project that the workers of the OT were used on was the West Wall. After World War II began, their primary work became reconstruction of necessary structures destroyed in the conflict. After Todt became Reich Minister for Armaments and Munitions, in 1940, the OT was increasingly involved with construction projects vital to the war effort. They followed the Army around Europe building oil tanks, industrial plants, rocket installations—including the V-1 ramps—Luftwaffe installations, the great U-boat pens in France, and the Atlantic Wall. Their repair work (bridges, roads, and factories) across Europe meant that they played a large role in exploiting the occupied countries. In the Balkans, the OT mined ores vital to the war effort, and shipped them to the Reich. The OT's greatest challenge came with the German invasion of the USSR, when the military construction battalions could not manage to build the necessary railways, bridges, and repair facilities.

The OT built all of Hitler's wartime Führer Headquarters. Following the failed bomb plot of July 20, 1944, Hitler's first thought was that it had been the work of a foreign OT worker, but he later apologized for this when it was discovered that it had actually been a serving German Army officer.[18]

A previously unpublished photograph of an unidentified German Army field marshal and Hitler (second from left) watching OT workers (at right). (HHA)

The Führer (left) receives the greetings of a small group of OT workers behind a chain link fence-enclosed work site on the grounds of FHQ Wolf's Lair, most likely during the major reconstruction phase of spring–summer 1944 in this very rare view. (HHA)

In five years, according to a British intelligence report, more than 1,400,000 men of the OT carried out the most impressive construction program in modern history.[19]

The OT wore brown uniforms and were under quasi-military discipline. From 1943, they wore insignia unique to the OT, and an armband bearing the phrase "ORG. TODT" was worn on the right arm above the swastika armband.

After February 1940, the OT expanded, recruiting large numbers of foreign workers, most of them non-German. However, the majority of OT work in the later war years was carried out by forced labor, often from the camps, and prisoners of war.

This raises some interesting questions: Had he lived, would Dr. Todt have also found himself in the dock of accused war criminals as did his successor, Professor Speer? Might he have resigned in protest against Hitler instead? We shall never know.

The OT reached its maximum size under Speer, who ran it much as Todt had. Later in the war, it became heavily involved in clearing air-raid damage in Germany. In the last years, the OT was involved in "anti-partisan" campaigns, which may in fact have been part of the Holocaust, and supervising slave labor: "They were technicians, slave-drivers, and—in some cases—murderers. Their technical ability was doubtless greater than their discipline. Complaints about corruption in the OT and other signs of organizational degeneration dramatically increased in 1944 as the Germans retreated on all fronts."[20]

However, the OT remained independent, even when in later 1944 it became the Front OT, armed to defend the Reich. After the war, the OT was broken up and banned by the Allies. Speer was convicted at Nuremberg of forced recruitment and employing slave labor.

Constructing Bridge by German wartime artist Handel-Mazzetti. (US Army Combat Art Collection)

The West Wall—which in September 1938 had been little more than a building site—now formed a deep defensive system, studded with 11,283 bunkers and gun emplacements...[21]

Lepage adds: "Militarily, the West Wall consisted of three major Army Defense Sectors. The lightly defended Northern Sector (or Aachen Sector) ran along the Dutch Limburg to Aachen. The Central Sector— the most important part of the West Wall—covered the area between Aachen, Trier, and Karlsrühe up to the Rhine. It concentrated its heaviest defense in the Pfalz, facing the French Maginot Line. The Southern, or Over Rhine, Sector covered the Upper Rhine region from Karlsrühe down to the Swiss border at Basel."[22]

Troop strengths had also been increased on the West Wall over the course of 1939: "The result was that on the Western Front, where in 1938 Adam had five regular and four reserve divisions, in 1939 [Wilhelm Ritter von] Leeb [commanding the Western Front army] had 12 regular and 10 reserve divisions as well as 15 *Landwehr* [militia] divisions"—quite a jump![23]

Hitler returned to inspect the works in the Saar territory just weeks before he invaded Poland, and also inspected troop maneuvers. "Reassured by the advanced condition of the structures intended to defend this section of the border, Hitler accepted 'reports on the readiness of the security forces as well as of the troops exercising in the zone.'"[24]

On November 23, 1939, Hitler received Todt, along with Hierl, Ley, and commanding generals of the Army and Luftwaffe in the west, at the Reich Chancellory to award them medals for their work on the West Wall: "On the front of the bronze medal ... a bunker was depicted above which there was a cross formed by a sword and a spade. The inscription on the back of the medal read: 'For labor in the defense of Germany.' The medal was worn on a brown band with a white rim."[25]

The medals were to recognize that the West Wall had fulfilled its designed military function in 1939, deterring the French Army from even attempting an invasion of the Third Reich while the bulk of its armies was off in the East subduing Poland.

Todt's work on the West Wall elevated his position with Hitler still further:

> Hitler took every opportunity to make clear to the Army generals that Todt was to take credit for the West Wall, arguing that if he had given this task to the Army alone, the fortifications would not have been ready for 10 years.
>
> With the West Wall, Todt proved that he was able to complete tasks that would be regarded as impossible by the standards of normal technical expertise. After this, Hitler's trust in Todt was matched by only a few in the higher ranks of the Nazi Party. Todt was Chairman of the Verein Deutscher Ingenieure [Association of German Engineers or VDI]...[26]

"People to Arms!" on a West Wall location, a Nazi propaganda slogan to inspire the Volkssturm (People's Storm or Militia) to fight in 1945. It failed. (CER)

Dr. Todt's 1943 Nazi film biography claimed, "Todt's influence at the West Wall caused a fundamental change in the military valuation of technology. He developed a totally different style of construction at the West Wall that was geared towards military purposes... A Swedish magazine wrote in an article about the 'artist with a thousand abilities,' and called him 'the greatest employer of all times, from the first minute of the war one of the most important members of the war machine. The spade had its place of honor next to the rifle.'"[27]

In November 1939, Göring appointed Todt to take overall responsibility for the entire construction sector. Though Todt hardly had a reputation for economy, he could at least be counted upon to ensure the absolute priority of the armaments effort.

Noted Gen. Siegfried Westphal in his postwar memoirs: "Hitler's goal in building the West Wall was without doubt achieved. The sprouting up of these fortifications along several hundred kilometers of frontier made a deep impression on the French and British politicians in the autumn of 1938."[28] German Army Gen. Hans von Mellinthin disagreed: "The troops were second class, badly equipped, and inadequately trained, and the defenses were far from being the impregnable fortifications pictured by our propaganda. The more I looked at our defenses, the less I could understand the completely passive outlook of the French."

The West Wall Under Attack

No one knew better than the Führer himself that the success of the Allied invasion of Northwestern Europe at Normandy meant the beginning of the end for his Third Reich, as the road to Berlin via France was far shorter than that of the Red Army coming from the east.

With the Nazi occupation of France in June 1940, the West Wall lay dormant for the next four years and two months, and its guns and troops were removed and sent to the newly built Atlantic Wall instead. The moment that Paris was retaken by the Allies on August 25, 1944, though, the Führer decided to reactivate the former West Wall defenses once more, to prevent an enemy wave from the west.

According to Max Domarus, Hitler reactivated the West Wall to face the oncoming Allies on August 30, 1944. "From the beginning, the Allied commanders realized the military insignificance of the eagerly erected West Wall. Although it would have been relatively easy for Allied pilots to bomb the construction site and disrupt the work, they made no such effort. Once the Western powers reached the 'Siegfried Line' in the spring of 1945, their tanks failed to fulfill the German propaganda promise that they would

Well, not quite, as the paper was dated September 16, 1944, and the fortifications held out until March 1945. (LC)

New York Post
Week-End Edition
FOUNDED 1801, VOLUME 143, NO. 238. COPYRIGHT, 1944, NEW YORK POST.
NEW YORK, SATURDAY, SEPTEMBER 16, 1944
5¢

Official: SIEGFRIED LINE COMPLETELY PIERCED

TODT'S RIVAL: THE MAGINOT LINE 1929–45

The linear defensive fortification on France's northeast frontier, the Maginot Line, was constructed under André Maginot in his last term as Minister of War (November 1929–January 1931). Work began in 1929, and was mainly completed in 1932, at a cost of three billion French francs. Maginot "became a symbol of the belief in elaborate fixed defenses that dominated French military thought before World War II, and stood in sharp contrast to German tactics that stressed rapid movement of tanks, planes, and motorized infantry…"[29]

It was built as a permanent defense against German attack, following the success that some French fortresses had had against German artillery during World War I. It was a huge advance on previous fortifications, with thicker concrete and heavier guns.

The Stalin Line, the Soviet Union's frontier fortifications against Poland, was built on the model of the Maginot Line. Czechoslovakia's frontier fortifications were also engineered by the French Army. When they fell to the Germans in October 1938 without a shot being fired, they gave the Germans valuable insights into the Maginot Line.

The defenders of the Maginot Line refused to surrender when attacked by the Germans in May 1940. However, the main thrust of the German Blitzkreig in 1940 outflanked the Maginot Line by attacking through Belgium and Holland. The portion of the Line in southern France held up to Italian attack in June 1940.

'plunge into the depths' of the supposedly impassable abyss. They crossed them by simply laying railroad tracks across the ditches and letting their tanks roll over them."[30]

In an order of the day in the fall of 1944, Army Field Marshal Gerd von Rundstedt proclaimed: "Soldiers of the West Front! I expect you to defend the sacred soil of Germany … to the very last! Heil the Führer!" but few of his compatriots believed that the now defunct Siegfried Line would be held for very long.

One of the more famous Allied commanders who agreed with them was Gen. George Smith Patton, Jr., who asserted on September 7, 1944 that, "Maybe there are 5,000—maybe 10,000—Nazi bastards in their concrete foxholes before the Third Army. If Ike stops holding Monty's hand and gives me the supplies, I'll go through the Siegfried Line like shit through a goose!" It proved to be a bold and foolish boast, however. As he also recalled, Patton stated that on December 6, 1944, "We arranged for a very heavy bombing attack on the Siegfried Line in the vicinity of Kaiserslautern. This was probably the most ambitious air blitz ever conceived. It was to consist of three successive days of attack, each one in considerable depth, and each day to consist of 1,000 heavy bombers… In order to prevent the enemy reoccupying the 4,000-yd strip thus evacuated, we planned to scatter tanks through the area immediately behind the bomb line… The capture of Metz and the Saar campaign began on November 8, 1944. On December 8th—that is, after one month's fighting—we had liberated 873 towns and over 1,600 square miles of ground."[31]

The overall campaign began on September 11, 1944 when a patrol of the US 5th Armored Division crossed the German border.[32] By the 15th, the Schill Line had been reached near Aachen by the 3d Armored Division also, and two days later, the American 12th Infantry Division arrived near Stolberg to find the "first substantial German reinforcements of the campaign."[33]

British Field Marshal Bernard Law Montgomery's ill-fated airborne/road-bound Operation *Market Garden* began on September 17, and by September 22, the US Army thrust against the Siegfried Line was halted because of a supply cut-off, mainly of gasoline.

The US drive resumed its forward momentum on October 2, 1944 to breach the West Wall north of Aachen, with an encirclement of that city in progress by the 8th. The American bombardment of Aachen began on the 11th, followed by an infantry assault on the 13th. The city's encirclement was completed on the 16th.

On October 21, 1944, Aachen became the first west German city to surrender to the Allies advancing from the west, and on November 2, the US 28th Division launched its attack on German forces in the "green hell" of the Hürtgen Forest. The battle for the area seesawed back and forth, including the commencement of Operation *Queen* with a heavy Allied aerial bombardment on November 16.

By December 7, 1944, the West Wall and Hürtgen Forest campaigns were effectively ground to a halt by the strong defenses,

The West Wall was designed to defend against the new machines of war, armed with many antitank guns, huge antitank minefields, walls and ditches, and the *Hickerhinernisse* – the dragon's teeth. This image well illustrates the *size* of the dragon's teeth. At the core, they had an iron triangle, around which a wooden frame was placed, into which the concrete was poured. When it dried, the frames were removed. (US Army Signal Corps)

and nine days later—on December 16th—Hitler launched his final surprise offensive of the war in the West, and the famous "Battle of the Bulge" began.

In January 1945, the stubborn German defense on the West Wall continued to hold out. Gen. Omar Nelson Bradley recalled: "We did not yet realize it, but the long, long pursuit was over. Along the entire length of the Siegfried Line, hastily organized German units were digging in to defend the Fatherland with utmost determination. Our great war of movement which had swept us 325 miles east of St. Lô in two months was, as historian Martin Blumenson put it, 'merging imperceptibly into a war of position.'"[34]

The Allies had an enormous military juggernaut to bring against the Wall: eight field armies totaling "about 2½ million men and half a million Allied vehicles in France."[35] Against them, around 63 numbered German divisions held the West Wall, though "Most of von Rundstedt's divisions were at no more than half strength and lacked guns and vehicles. Some were composed of hastily drafted old men and young boys with only rudimentary training, but what the Germans lacked in manpower and hardware, they made up for in *esprit*."[36]

"In three weeks, Patton slugged his way forward some 35–45 miles to the Saar River, crossed it, but was stopped dead at the Siegfried Line."[37] Patton's drive to the Siegfried Line cost 27,000 casualties.

Eventually, in March 1945, an attack of three corps broke through the lines, and, realizing they were trapped, Field Marshal Albert Kesslering authorized a withdrawal.[38] Thus had the mighty, once vaunted, West Wall/Siegfried Line fallen at last.

This famous image of the Siegfried Line shows men of the 39th Infantry Regiment (9th US Infantry Division) passing through the Siegfried Line near Roetgen. In the immediate foreground are iron girders that were placed in the recesses of the tragesperre to block the road. To the sides are dragon's teeth and improvised antitank obstacles which are simply wooden posts rammed into the ground. (US Army Signal Corps)

TODT AND THE WEST WALL

Hitler's eulogy at Todt's funeral in 1942 covered Todt's role in the defenses of the west and subsequent tasks in detail:

In the Four Year Plan, he was given a special position as Inspector General for special projects... War danger began to gather about Germany... I was obliged to make provision for the defense of the Reich on a large scale, and as soon as possible. I had conceived the plan of erecting a fortification opposite the Maginot Line ... which was to protect the vitally important western portion of the Reich against any attack....

There was only one man who was in a position to solve this technical engineering problem, unique in the history of the world—and to solve it, indeed, in the shortest possible time.

On May 28, 1938, I made known my resolve to the Army and the Luftwaffe, I entrusted the Inspector General for Construction Dr. Todt with the responsibility and supervision of the construction of the largest part by far of this gigantic new work—in cooperation with the proper military authorities ... that as early as September 1938 at the latest, at least 5,000 concrete and steel positions would have to be ready or usable.

The whole program was planned with a total of 12,000 units ... which in a year and a half increased to 23,000 units. The present war experiences have confirmed our conviction that no power in the world could succeed in breaking through this most gigantic defense zone of all time!

This marvel is, in its technical plan and purely organizational measures of construction, as well as in the technical building itself, for all time associated with the name of Dr. Todt.

The war which broke out presented new problems to this greatest organizer of modern times. A system of great roads for deploying troops had to be built up in those regions of the Reich in the shortest possible time, which previously had been very much neglected. Thousands and more thousands of kilometers of roads were either newly built or widened, provided with a hard surface and made dust proof.

When the fighting finally began, units called into being by this unique talent for organization marched behind and forward with the troops, removed obstacles, rebuilt destroyed bridges, improved roads, erected everywhere new junctures over valleys, ravines, rivers, canals, and thus complemented in an indispensable way the engineering troops ... and thereby could enter more actively into the fighting.

The victory in Norway and the victory in the west brought new tasks. After former Party comrade Todt had been named to the Reich's Ministership for Armaments and Munitions—and thereby had to organize and lead a new, truly formidable sphere—there came in addition the task of protecting against enemy attacks through the construction of new, powerful fortifications.

Now, not much of the West Wall is left, most of it having been destroyed after 1945. This was not an easy task, as Lepage explains:

The destruction of the West Wall was a very costly business, as the blowing up and clearing of a bunker was estimated to cost 8–60,000 Deutsche Marks. Today, all one can see of Fritz Todt's *chef-d'oeuvre* are a few units in the eastern side of the former line, a few undamaged small bunkers in the hinterland, and a few sections of antitank dragon's teeth, heaped-up ruins submerged along the vegetation, and half-buried in the open countryside.[39]

As for Dr. Fritz Todt, in March 1940 he approached the final great challenge of his career, as we shall now see.

CHAPTER FOUR

MINISTER OF ARMAMENTS AND MUNITIONS, 1940-42

"Give us the weapons for the Front"
German wartime Nazi propaganda poster

"Give us the weapons for the Front."
(Courtesy George Peterson, National Capital Historic Sales, Springfield, VA)

On May 21, 1935, the Führer delivered in the Reichstag in Berlin his second "peace speech"—reiterating many of the points he had made in his address of May 17, 1933. The 1935 version was meant to allay European and global fears of German rearmament, announced the previous year on March 15, Heroes' Memorial Day, in Berlin, the same Saturday on which Herr Hitler had also announced the return of the German military draft for males.

During this major address, the Reich Chancellor made mention of Republican Germany's disarmament in accord with the Treaty of Versailles that it had signed with the victorious Allied powers on June 28, 1919:

Germany performed the obligations imposed ... with nothing short of zealousness: financially, up to the complete collapse of its finances; economically, up to the total destruction of its economy; militarily, up to a complete lack of defenses. I ... repeat here in general terms the facts of Germany's performance of the treaties which are contested by no one.

The following were destroyed in the Army: 1) 59,000 guns and barrels; 2) 130,000 machine guns; 3) 31,000 trench mortars and barrels; 4) 6,000,000 rifles and carbines; 5) 243,000 machine gun barrels; 6) 28,000 gun carriages; 7) 4,390 trench mortar carriages; 8) 38,750,000 shells; 9) 16,500,000 hand grenades and rifle grenades; 10) 60,400,000 live fuses; 11) 41,000,000 rounds of small arms ammunition; 12) 335,000 tons of shell cases; 13) 43,515 tons of cartridge cases; 14) 37, 600 tons of gunpowder; 15) 79,000 ammunition gauges; 16) 212,000 telephone sets; 17) 1,072 flamethrowers, etc.

A German wartime industrial propaganda poster proclaims, "Front and Homeland: the guarantee of victory." It shows a cadet, housewife, soldier, and industrial worker, united under the banner of the swastika. (CER)

The poster shown here was for the German Pharmacist/Druggists Day at Essen of June 23–27, 1937, with an industrial plant in the background. (CER)

Further destroyed were sledges, mobile workshops, antiaircraft vehicles, limbers, steel helmets, gas masks, machines of the former war industry, and rifle barrels. Further destroyed in the air were: 1) 15,714 fighter planes and bombers; and 2) 27,757 aircraft engines.

At sea, the following were destroyed: 26 capital ships, four coastal tankers, four battle cruisers, 1 light cruiser, 21 training ships and special ships, 83 torpedo boats, and 315 submarines.

Also destroyed were motor vehicles of all types, chemical warfare and, in part, anti-gas defense equipment, propellants, explosives, searchlights, sighting devices, range finders and sound rangers, optical instruments of all kinds, harnesses, etc.; all airplane and airship hangars, etc.[1]

This, then, was the military-industrial situation that the new German Reich Chancellor had found upon taking office on January 30, 1933. A year before taking office, on January 27, 1932, Hitler had addressed German industrialists at the Industry Club at Düsseldorf, announcing his intentions to rearm Germany when he took office.[2]

A previously unpublished German period woodcut entitled *The Worker* by artist Georg Sluyterman von Langeweyde. The text translates as: "At this time, there will be only one aristocracy: the aristocracy of 'work, labor, toil, and employment.'" (US Army Combat Art Collection, courtesy Curator Renée Klish, Fort Leslie J. McNair, Washington, DC)

On the evening of February 3, 1933—his fourth day in office—the new Reich Chancellor spoke again on the topic, this time to the assembled top generals and admirals of Weimar Republican Germany, at the invitation of the Army Commander in Chief, Gen. Kurt von Hammerstein. Hitler spoke for two hours, and gave the martial leaders of his planned Third Reich what they most wanted to hear: a recreated war machine in time of peace with an exciting vista of promotions, commands, increased salaries, and other emoluments. Most, but not all of them, were sold.

On March 17, 1935, as Führer and Reich Chancellor, Hitler unilaterally abrogated the 1919 Treaty of Versailles by announcing the return of German universal military conscription, and, with that, a vast expansion of the peacetime German Army.

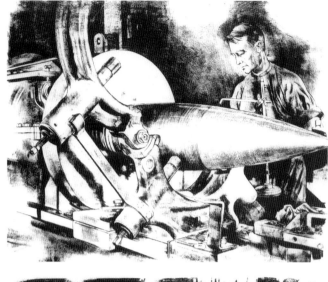

The new German Luftwaffe was announced as already in being. Later that year, the Reich government signed the Anglo-German Naval Treaty enabling the Kriegsmarine (German Navy) to expand as well.

In the summer of 1936, the Führer took another giant stride toward war by trying to make the Third Reich self-sufficient via the new Four Year Plan headed by Prime Minister of Prussia and Aviation Minister

Previously unpublished wartime lithographs by artist Josef Arens (clockwise from top left): *Shaping German Shell on Lathe*; *Gutting Gun Barrel on Large Lathe*; *Pouring Iron Casting*; *Steel Mill Operation*. (US Army Combat Art Collection)

GUSTAV KRUPP

Nazi Germany's premier heavy artillery producer was "The Cannon King," Gustav Krupp von Bohlen und Halbach (1870–1950). Born the son of a German diplomat at The Hague on August 7, 1870, Gustav Krupp made his family fortune in the coal and iron fields of Pennsylvania before returning home to the new, united, Imperial Reich of the Kaisers. Serving as a German Foreign Office diplomat, Krupp had postings at Washington, DC, Peking, and The Vatican in Rome.

He married Bertha Krupp in 1906. Bertha had inherited her family's company four years previously, at the age of 16, after her father committed suicide. According to one account, "The marriage had been arranged because it was unthinkable at the time for such a large company to be run by a woman. The Kaiser announced at the wedding that Gustav von Bohlen und Halbach could now use the name Krupp, and he became company chairman of the board in 1909." Krupp ran the firm from 1909 to 1941. He was typical of the German heavy industrialists with whom Hitler, Göring, and both Drs. Todt and Speer dealt before and during the war.

During World War I, a 94-ton howitzer was named for his wife, the famed *Big Bertha* that bombarded Paris. He finally agreed to back the Nazis in 1933 after they promised to destroy the Socialist trade unions, and to vastly increase the size of the German Armed Forces. Krupp later became chairman of the prestigious Association of German Industrialists and the Adolf Hitler Fund. On August 30, 1940, Krupp became the first recipient of the Nazi "Pioneer of Labor" decoration. Krupp received this badge from the Führer personally, as well as the coveted War Merit Cross, for his 70th birthday at Essen in the German industrial Ruhr.

Hermann Göring. This plan took Nazi Germany into early 1940, and midway into the first full year of World War II.

On March 17, 1940, Hitler named as his first Minister of Armaments and Munitions none other than his eminently successful road builder, Dr. Fritz Todt, who until then had been busy constructing roads and bridges via his engineers.

Todt also ran the Head Office for Technology, and all the major technical tasks of the Third Reich's war effort were thus concentrated in his hands. In addition, Dr. Todt was responsible for road building in all the territories occupied by the Germans from the northernmost tip of Norway to southern France.

His workload was thus immense, as noted in his 1943 Nazi biography:

For many years, he lived in two simple rooms on the garden side of the Service Building, Pariser Platz 3. When he got to the apartment in the evening, he brought some of the most difficult work with him. Even on the days he spent with his family in Munich, or in the country house on the Hintersee, Dr. Todt had meetings to attend, or inspections. He wrote in 1934, "Mostly I come to Munich when it gets too hectic in Berlin, or when it is important for me to be in Munich. I plan nothing on my 'quiet day', and the result is there are so many things to do that—before you know it!—it's time for the train to Berlin." Hardly a second co-worker traveled as much as he. The simultaneous start of the autobahn in many areas of the Reich, the construction of the West Wall, the protection of the Atlantic coast, the campaign against Russia, and, finally, the care of the armaments industry, demanded a heavy travel schedule, which in eight years came to 700,000km by vehicle and 400,000km by air.[3]

In 1940, Hitler was determined to implement a massive attack in the west because he thought this was the only way that Germany could win the war. As Tooze explains:

> Regardless of how intensively its home front was mobilized, Germany would lose a protracted war, because the combined might of its enemies was simply overwhelming. Germany therefore needed to concentrate all its resources on striking a decisive blow at the earliest possible opportunity.
>
> If this meant temporarily sacrificing the needs of civilian consumption, so be it. As Hitler commented to the chief of Army procurement, Gen. Karl Becker, in early November 1939, "One cannot win the war against England with cookers and washing machines."

A previously unpublished German wartime watercolor by artist Bregle entitled "*German Göring*" *Defense Construction*, most likely of the HG Works at Salzgitter, Austria. (US Army Combat Art Collection)

A previously unpublished German wartime painting by artist Richard Gessner entitled *Smelting Works.* (US Army Combat Art Collection)

In pursuit of victory, the question that preoccupied the key players in Berlin was not how to balance the needs of the war effort and the civilian economy. The question was how best to organize the economy for total war.

The military-economic staffs of the OKW (High Command of the Armed Forces), the Ministry of Economic Affairs (Göring and Dr. Walther Funk), and the Reichnährstand (the national agricultural organization) were all haunted by memories of 1914–18...[4]

During World War I, Imperial Germany had been largely defeated by the sea blockade of all German and Central Powers ports by the British Royal Navy. This, in turn, had been instrumental in bringing about what the Nazi Party—and chiefly, Hitler himself—most feared: an internal revolution as a result of civilian domestic discontent that would overthrow the regime at home, just as it had the German Kaiser, Emperor Wilhelm II, in late 1918. This haunting image—spurred on by the defeated marshals and generals of the German Army then—was always uppermost in Hitler's political thinking.

As not only Head of Government, but also political Chief of State as well, the Führer—like his foes Stalin and Roosevelt (but not Churchill nor Mussolini)—had to consider the entire, overall, strategic global view of the war; and in the Third Reich, he and he alone had all the varying threads of the information highway leading to him. Only he knew the whole picture at any given moment.

Thus, while his underlings increasingly demanded total war on its military-economic merits alone, it was the Führer who had to balance these demands against that of keeping the people at home as happy as he could, waging a relatively painless war. Professor Tooze observed:

Nobody in 1939 expected Germany to be able to last as long as World War I. The Third Reich's deficiencies of foreign exchange and raw material stocks were too severe, but under the direction of State Secretary Herbert Backe, the Reichsnährstand was preparing for a three-year war.

General [Georg] Thomas of the OKW and his collaborators in the Ministry of Economic Affairs believed that with careful husbanding, Germany's stocks of industrial raw materials could be stretched over a similar period of time.[5]

Thus, the great debate between October 1939 and May 1940 was how to use Nazi Germany's existing resources. The Hitler–Göring–Keitel–Todt faction wanted to stake everything on a single sharp blow in the west, while that of Thomas–Backe and others wanted to dole out those same stocks to survive the feared long war that Hitler also thought wouldn't end in a German victory until 1950, a secret he confided only to his master builder, Professor Speer.

The rationality of the two opposing arguments aside, Hitler held the cards politically, since he outranked everyone else, while the German military stonewalled the launching of the western campaign from fall 1939 (when the Führer urgently wanted it) until it finally began, after many postponements, on May 10, 1940.

Professor Tooze testified, "On March 24th [1940], General Thomas noted the following conversation with Fritz Todt: 'Führer has again emphasized energetically that everything is to be done so that the war can be ended with a great military victory in 1940. From 1941 onwards, time works against us.'"[6] Far from underestimating the industrial potential of the US in the war later on, both Hitler and Göring fully recognized what had happened during 1917–18, no matter how much they might belittle the Americans to their respective entourages, and no two men also knew better than they the relative strengths of the US and Greater Germany.

The real armaments crisis confronting them in late 1939, however, was a lack of needed ammunition above all else. Professor Tooze asserted: "The fundamental question was one of resources. After aircraft production, supplying the enormous volumes of ammunition demanded by modern warfare was by far the largest industrial challenge facing the German economy in World War II."[7]

It was partly for this reason that the SS killers used gas to kill their victims during the middle and last phases of the Holocaust, as a means of saving precious ammunition.

"The main suppliers of ammunition were the giants of German heavy industry, firms such as United Steel Works, Krupp, Klöckner or the Reich Works Hermann Göring,"[8] and Hitler began demanding more and more production in preparation for the coming blow against the West.

A previously unpublished wartime portrait of Nazi Reich Minister for the Occupied Eastern Territories architect Dr. Alfred Rosenberg (1893–1946) by artist Pitthan. He wears the Golden Party Badge on his left breast pocket and the ribbon of the Blood Order on his right, as well as a Party leader's armband. Dr. Rosenberg was one of Dr. Todt's Cabinet colleagues, then Prof. Speer's. Among many other posts, Rosenberg was a general in both the SA and SS. Born January 12, 1893—the same date as Hermann Göring—to Baltic German parents at Reval, Estonia in the then Russian Empire, Dr. Rosenberg studied architecture at the Riga Polytechnic Institute, and engineering at Moscow University, completing his Ph.D studies in 1917. His ministry deputy was Alfred Meyer. Had the Nazis won World War II, Dr. Rosenberg would have held one of their premier postwar jobs as the major domo of occupied Russia, with expected turf battles with Goring, Reichsführer-SS Himmler, and Dr. Speer in the exploitation of the vast new territories. (US Army Combat Art Collection)

Reich Chancellor Hitler and two of his top generals in East Prussia in 1933, before the Nazi eagle and swastika emblem had been added to their peaked military caps and tunics. At left is the then German Defense Minister Gen. Werner von Blomberg (whom Hitler made a field marshal on April 20, 1936), and at center (to Hitler's left) is Army Gen. Walther von Brauchitsch, whom the Fuhrer created field marshal in the summer of 1940 after the fall of France. Von Blomberg was toppled from his post in a sex scandal early in 1938, when Hitler took the post of War Minister himself. Von Brauchitsch served as Commander in Chief of the Army 1938–1941, when Hitler fired him, and took over that post as well. Both Drs. Todt and Speer were colleagues of these two eminent soldiers. Third from right is the Nazi Gauleiter of East Prussia, Erich Koch. (Previously unpublished photograph, HHA)

"Far from lacking clear priorities," noted Professor Tooze, "the German industrial war effort was dominated by only two components: aircraft and ammunition... Between them, these two items claimed more than two-thirds of the resources committed to all armaments production in the first 10 months of the war," or until June 1940, during the defeat of France. Everything else that needed to be produced—such as tanks, artillery, and ships—had to be made from the remaining third.[9]

During the winter of 1940, the Army—which was then responsible for ammunition production—began to see an upturn in production, but it wasn't enough to satisfy Hitler, as indicated in the following February 1940 message sent from OKW chief Col. Gen. Wilhelm Keitel to his nominal subordinate, Army C-in-C Col. Gen. Walther von Brauchitsch:

The Führer and Supreme Commander of the Wehrmacht, after taking note of the last reported monthly production of weapons and ammunition [January 1940], has expressed to me his dissatisfaction with the low level of performance achieved since the beginning of the war... The modest production of ammunition in January in the most important calibers leaves little hope that the Führer Order will be satisfied as of April 1, 1940. The Führer stated that it is necessary to reorganize the (Army) procurement system for ammunition production.

Truly, this had ominous overtones for the Army, and the question immediately arose: whom did the Reich Chancellor have in mind to achieve the demanded rise in ammunition production? Doubtless, they feared that this key post would fall to one of their main antagonists in

Nazi Germany, Field Marshal Göring, and he thought similarly, but the real answer wasn't long in coming, and surprised them all, as Professor Tooze confirmed:

> The man who was to replace the Army bureaucrats in charge of ammunition production—the most important aspect of armaments production other than aircraft—was Hitler's favorite miracle worker, Fritz Todt. It was Todt who had made real the Führer's vision of a national system of autobahns.
>
> In 1938, it was Todt who on the West Wall had demonstrated that the initiative of the politically committed engineer—harnessing the energies of private business—could deliver what the lumbering military bureaucracy apparently could not. And it was Todt who was duly appointed by Hitler on March 17, 1940 to head a new Ministry for Ammunition.
>
> This was a calculated snub to the officers of the Army procurement office, and it was bitterly felt... On the night of April 8th, Becker's efforts to salvage the Army's position were sabotaged by Erich Müller, aka "Cannon" Müller, the chief weapons designer of Krupp, a man who made a career out of indulging Hitler's whimsical ideas for oversized artillery pieces.
>
> Seizing the opportunity for an impromptu interview on the Führer's private train, Müller impressed upon Hitler that German industry wanted to see the new Ammunition Ministry headed not by a soldier, but by Todt.[10]

And thus it came to pass that a highway engineer found himself as a new cabinet minister in the Third Reich, holding one of Nazi Germany's most important governmental portfolios! Hitler's famous intuition and

Seen as a lackey to the Führer he idolized, here Field Marshal Keitel (left) makes a point to Hitler outside Bunker #10—Kasino #1—at FHQ Wolfsschanze (Wolf's Lair). The Führer distrusted Keitel, however, and maliciously told Dr. Goebbels that Keitel had "The brains of a movie theater usher." Keitel worked closely with both Drs. Todt and Speer throughout World War II, and was hanged at Nuremberg as a convicted war criminal. (HHA)

U-BOAT BUNKERS

After the fall of Norway, France, and the Low Countries in 1940, the Germans acquired a large number of bases for their U-boats, many of them bases they had used in World War I. New U-boat bases were established in Norway and France. These were very useful to Adm. Karl Dönitz, but were also within easier reach of Allied bombing raids. The Germans now used their World War I experience of concrete bunkers at Bruges, which had remained mainly undamaged despite Allied bombing raids, and built new bunkers on similar lines at all major U-boat bases in France and Norway. These were bombproof, but as the RAF developed heavier bombs later in the war, they were strengthened further. The concrete roofs were sometimes over 7m thick, strengthened with steel. The U-boat bunkers were constructed by the OT, under Todt and later Speer. French laborers also worked on the new pens, as did forced labor: interned Spanish Communists, Soviet prisoners of war, and concentration camp inmates. The workers did 12-hour shifts, with work going on through the day and night, stopping only when air-raid warnings sounded.

Previously unpublished German wartime artwork of a U-boat on patrol on the surface. (US Army Combat Art Collection)

impetuosity had demonstrated itself once again, but doubtless he had had Todt in mind for quite some time. The Führer seemed to accept Müller's suggestion in order to placate German heavy industry, but Todt was also one of his very top men, respected by everyone in the hierarchy, both military and civilian. It was an inspired choice.

According to van der Vat: "Todt was not enthusiastic about this new post, because he felt he lacked the technical expertise; he was a civil engineer, not a gunsmith. There was a cautious welcome from the relevant industries, which had observed how he had created huge new opportunities for construction firms and materials suppliers by means of fundamental reorganization of working practices, affecting management at least as much as workers."[11]

The Nazi Party victory of the Hitler–Göring–Todt faction over the Army is effectively analyzed by Professor Tooze:

> The war … began with a convulsive rearrangement of power and influence within the Army's segment of the armaments economy. The soldiers were the losers. After Becker's suicide, the Army procurement office never recovered its authority. In addition, the ambition of the Wehrmacht military-economic organization to exercise overarching control over the entire armaments economy was dealt a fatal blow by Fritz Todt's emergence as an independent force in armaments politics.

The winners in this power struggle, apart from Fritz Todt and the Nazi Party, were the leaders of German industry, who had rallied around the new Ammunition Minister in April 1940...

The combination of military victory and surging production statistics was an intoxicating propaganda cocktail, which Fritz Todt and his backers in the Nazi Party exploited to full effect. In his victory speech of the summer of 1940, Hitler gave credit for the armaments effort entirely to Fritz Todt, pointedly ignoring both the Army's own procurement officers and General Georg Thomas of the OKW. This rhetoric accurately reflected the outcome of the political battle, but it had little to do with the realities of production.[12]

Hitler's words to the German Reichstag on July 19, 1940, were reported by Max Domarus:

In the wake of the armies followed the commandos of the Todt Organization, of the Reich Labor Service, and of the NSKK, and these also helped to repair roads [and] bridges, as well as to restore order to traffic... Party Comrade Hierl has been the founder and leader of the RAD. Party Comrade Ley is the guarantor of the behavior of the German worker. Party Comrade and Reich Minister Major General Todt is the organizer of the production of armament and ammunition, and has gained eternal merit as a master builder of the fortified front in the West.[13]

German Army horse-drawn machine guns (left) and field artillery (right, center) at one of the annual Party Days at Nuremberg's Zeppelinwiese (Zeppelin Field). Horses played a prominent part in all German military campaigns throughout the war, from Poland 1939 through Normandy 1944, despite Hitler's ardent wish to disband all of the Reich's cavalry corps as an anachronism. They proved their worth especially on the Eastern Front. (HHA)

Göring (third from left), Hitler (fifth from left), and Dr. Ferdinand Porsche (seventh from left) at one of the annual Berlin International Automobile and Motorcycle Shows before the war. Note the Ford Motor Company wall sign at left, and the Krupp vehicle at right in this previously unpublished photo. (HGA)

The repetitive mention of the moniker "Party Comrade" was the Nazi Führer playing Nazi politics from the tribune of the nation's top legislative body, and Dr. Todt, too, also knew how to wage high politics, as noted by Professor Tooze: "If Todt did not owe his appointment to a business conspiracy, it cannot be denied that—once in office—he actively sought an alliance with German industry, and that the Reich Group for Industry responded with enthusiasm."[14] Dr. Speer would repeat this successful Party-private sector gambit once again in 1942. After his appointment, Todt conferred with various heads of industry:

> Todt kept General Thomas of the OKW waiting whilst he conferred with Albert Pietzsch of the RwK [Reichswirtschaftskammer, Reich Economic Chamber] and Wilhelm Zangen of the RgI [Reichsgruppe Industrie, Reich Group for Industry]. He also had meetings with Philip Kessler of the Fachgemeinschaft Eisen und Metallindustrie [Specialty Group of the Iron and Metal Industry], Borbet of the Bochumer Verein [Union], Karl Lange of Business Group engineering, and Rudolf Bingens of Siemens.[15]

By the end of one week in office, Todt had committed to setting up an industrial council, and had been given a list of RgI's key priorities.[16] Todt would spend the next few months trying to implement measures to satisfy the RgI. The most important of these was a modified system of pricing for ammunition orders, and a decentralization of the distribution of ammunition orders, which worked by having regional ammunition committees that coordinated the producers interested in each caliber of ammunition. Army procurement officers issued orders to these committees, and the industrialists would assign the contracts to the most suitable producers.[17] Coordination between regions was established by an ammunition council in Berlin which liaised between

Todt's ministry, and the metal working industries. Dan van der Vat noted the impact of Todt's actions:

> Todt's approach to the armaments industry alarmed the Wehrmacht because he began to impose a pragmatic compromise between military specifications and industrial capacity and potential, instead of allowing Wehrmacht requirements to dictate the production of an endless proliferation of short-run items, models, types, or marks.
>
> Todt also set up quasi-autonomous working groups of companies and committees to coordinate supply and production in the main sectors of the armaments industry. The underlying principle was industrial "self-responsibility"—the autonomous responsibility of industry for how it executed its assigned tasks. So long as a given industrial sector met its targets, it was left to itself, and insulated from interference by generals, bureaucrats, and even Party bosses.[18]

This, in effect, was the very same "self-responsibility of industry" and its connected "rings" of various segments for which Professor Speer was later given credit by both himself and postwar historians as having effected the 1942–44 German armaments production "miracle." The question arises: what happened, then, to Dr. Todt's previous initiative?

The role of the ammunition committees expanded to cover overseeing the production capacity of all factories in the region. Todt then gave them the power to issue orders so that production targets were met, and to deny orders to firms if they lacked the necessary labor and resources. Professor Tooze describes the changes thus:

Dr. Fritz Todt (left, center) and Göring (center) walking together during the early part of the war. At far right is Göring's FHQ liaison officer, Gen. Karl-Heinrich Bodenschatz. Göring's Luftwaffe swagger-stick baton is still missing. (HGA)

Members of German builders and other Nazi groups congratulate Reich Marshal Göring (right) at his country home, Karinhall, outside Berlin, on his 49th birthday, January 12, 1942. Seen here in this previously unpublished photograph is Alfried Krupp von Bohlen und Halbach, third from right. Alfried Krupp ran the Krupp Works during 1941–45, when his father Gustav retired due to ill health. He also stood trial and was convicted as a war criminal at Nuremberg in his stead. He spent 1948–51 in prison, when he was amnestied, and regained control of the company in 1953. (HGA)

By May 1940, Todt had moved from a strictly regional system to set up national subcommittees in Berlin for all major types of ammunition. In the summer, he followed this up by announcing the formation of a national committee for tank production, which began to meet from the autumn of 1940 under the chairmanship of Walter Rohland of the Edelstahlwerke.[19]

This process culminated in a national committee for the production of guns and artillery headed by Erich Müller from Krupp. However, once the ammunition crisis of 1940 had passed, Todt's bid to take control of the armaments effort faltered as a result of bureaucratic in-fighting. Tensions grew between Todt, the OKW's military-economic office, and the armed services' procurement offices. Göring's position as head of the Four Year Plan, and head of the Luftwaffe further complicated matters, and the head of the civilian economic administration, Walther Funk, also got involved. Accusations of incompetence and inefficiency were constant. Speer later remembered a letter from Todt which reveals something of his feelings about the tensions between the different factions:

In January 1941, when I was having difficulties with Bormann and Giesler, Todt wrote me an unusually candid letter which revealed his own resigned approach to the working methods of the National Socialist leadership: "Perhaps my own experiences and bitter disappointments with all the men with whom I should actually be cooperating might be of help to you, enabling you to regard your experience as conditioned by the times, and perhaps the point of view which I have gradually arrived at after much struggle might help you somewhat psychologically, for I have concluded that in the course of

German tanks in the USSR, 1941. (*Signal* magazine, George Peterson, National Capital Historic Sales, Springfield, VA)

such events ... every activity meets with opposition, everyone who acts has his rivals and, unfortunately, his opponents also. But not because people want to be opponents, rather because the tasks and relationships force different people to take different points of view. Perhaps, being young, you have quickly discovered how to cut through all such bother, while I only brood over it.[20]

In his memoirs, Speer complained about the inefficiency and under-mobilization of the German war economy of 1940–41, but Tooze contests that the bureaucratic wranglings should be disregarded, as in fact, "The politics of the German war record of industrial production between June 1940 and June 1941 in fact bears the unmistakable imprint of strategic design"—by Hitler himself.[21] Armament production and economic policy in this period were part of Hitler's strategic war plan, which did produce a further mobilization of the German economy, though not enough to defeat the USSR.

The invasion of the USSR was coupled with a tacit pact of industrial production agreement between Reich Marshal Göring and his Luftwaffe subordinate, Air Force Maj. Gen. Dr. Fritz Todt: "The floodgates in Luftwaffe planning finally opened in the summer of 1941 with the completion of the Army's *Barbarossa* (Red Beard) program,

and the long-awaited decision to shift priority to the air war. In June 1941, the Air Ministry proposed a doubling of output to 20,000 aircraft per year over the following three years," i.e. to 1944, when—had that happened—Göring might have been able to win the aerial war against the Allied air forces.[22]

Germany needed to increase aircraft production by 150%, to around 3,000 planes a month, and so Göring's staff and Todt agreed that resources would be reallocated to the Luftwaffe, with Todt overseeing the identification of spare capacity, and ensuring that Army contractors were kept employed.

Despite this, Hitler later reversed the Göring–Todt pact, restoring industrial production capacity priority to the Army, in order to better defeat the Red Army on the ground on the Eastern Front.

Todt's workload increased further, as Hitler made him Inspector General for Water and Energy in July 1941. This would be his last appointment. According to van der Vat, "this was welcome because it was a rational extension of existing OT responsibilities and suited to his engineering skills, although even he could not make much of it in the six months he was to hold the title."[23] Also, following Hitler's invasion of Soviet Russia on June 22, 1941, the OT reconstructed miles of Russian narrow-gauge railways as well, altering them in the process to the standard German gauge, or width, as well as establishing depots adjacent to the Moscow front.

In the fall of 1941, Todt became personally aware of the "Judeocide" and the genocide of other groups for the first time officially—if, indeed, he did not already know about it:

> On October 17th—in front of Armaments Minister Fritz Todt and Gauleiter Fritz Sauckel—Hitler presented a veritable panorama of the future of the conquered East. The Slav inhabitants were to be treated like "Red Indians"... Those who were resettled would be retained. The population of the cities was to be starved to death. All destructive Jewish elements were to be exterminated immediately... The Third Reich was already committed to a program of million-fold murder that aimed at nothing less than the demographic reconstruction of Eastern Europe.[24]

Thus, like Professor Speer later, Dr. Todt *knew*.

There was even worse news for the Third Reich militarily that fall, as the rains over Russia turned the ground to mud, soon to be followed by ice and snow. Charged with extending his much vaunted autobahns to the very ends of the projected Nazi Eastern empire to the foot of the far-off Ural Mountains in Soviet Asia, Dr. Todt was all too aware of the difficult task facing the German war machine in accomplishing such a gigantic, multi-pronged task: conquering, garrisoning, resettling (with Germans), and pacifying the vast reaches so grandly spread before him.

ENGINEERING AND MILITARY ENGINEERING

Engineers have been crucial to powerful leaders since the time of the Ancient Egyptians. The monumental works of Ancient Egypt, and the achievements of the Romans show that "the principles of civil, mechanical, and hydraulic engineering were well understood before the Christian era."[25] English Army records from the 14th century note the number of engineers borne on the strength of the ordnance. These engineers did not only direct weapons, but also undertook the design and construction of fortifications, roads, and bridges.

The real birth of civil engineering dates to the early 17th century, when neglect of the rivers of northern Italy caused many disastrous floods. Through consultation and experimentation, a group of men emerged who could deal with hydraulic works and mechanics. Their work extended to cover the design and construction of bridges, roads, and general machinery. As their work was similar to that of military engineers, they adopted the same title.

Scientific advancement over the centuries has led to diversity within civil engineering, so that mechanical, electrical, aero, agricultural, chemical, hydroelectrical, and metallurgical engineering "are now to all intents and purposes separate professions."[26] Engineering is so integral to modern society that "civilization in the modern sense of the word and engineering may be said to be synonymous."

It was only relatively recently, in the last two centuries, that engineering education began to be more organized and systematic. The US saw a huge rise in the number of technical colleges following the American Civil War. In the mid-1880s, Germany was developing its industrial education, reorganizing existing technical schools, and establishing new ones. By the end of the 19th century, Germany was a world leader in engineering education, rating scientific investigation and technical education highly.[27] This was the environment within which Fritz Todt began his studies.

Military engineering underwent a revolution between World War I and World War II. In World War I, engineers worked with pick and shovel, explosives, locally sourced material, and conscripted labor. The development of mobile earth-moving equipment such as bulldozers; of rapid bridging equipment; of mass production of antitank and antipersonnel mines; and of prefabricated surfacing materials dramatically increased the capacity of engineers to affect military operations.[28]

Also, an army's need for engineers increased as many roads and bridges were not suitable for large numbers of heavy military vehicles. And as the scale of war increased, so logistics became ever more complex, and the reliance on engineers increased: "Engineer units constituted an average of about 12% of most field armies in the Second World War."[29] These were split between those who worked with units in the field, and those trained and equipped to carry out major engineering tasks throughout the theater. In addition "there were specialized engineering organizations at the national level, set up to undertake strategic tasks, [such as] the Todt Organization."[30]

Oil painting by O. Anton entitled *Assault Crossing of Rhine River by SS Troops*. The MG 34 machine-gun has its bipod folded, and was available in both light and middleweight versions. Its drum magazine held 50 rounds, a double drum 75. Both the Waffen (Armed SS) and the regular German Army used engineering troops such as these to ford rivers, build bridges and roads, and other field structures and fortifications. Engineering is the science of striking a balance between cost and weight, safety and strength. (US Army Combat Art Center, Washington, DC, courtesy of Marylou Gjernes)

General Thomas had had economic misgivings—as had Göring—about the chances of conquering the Soviet Union from the very beginning, and in December 1941, told OKW's chief, Field Marshal Keitel, that the Eastern Army might not have enough fuel to propel the projected offensives against the Red Army set for summer 1942. Keitel told Hitler, and thus the stage was set for the eventual appointment of Professor Speer to succeed Dr. Todt, who—apparently—had failed.

"Even as ardent a Nazi such as Fritz Todt ... was under no illusions about Germany's situation," noted Professor Tooze. "According to both Walter Rohland—head of the Main Committee for tank production—and Hans Kehrl, Todt had harbored serious reservations about the Russian campaign from an early date."[31] According to Dr. Speer in his 1971 interview with journalist Eric Norden in *Playboy* magazine:

> One man who tried to throw cold water on the leadership's delusions of omnipotence was Dr. Fritz Todt, the Minister of Armaments and Munitions, with whom I worked closely in my capacity as chief of armaments construction. He believed that our troops were neither physically nor psychologically prepared for the rigors of the Eastern Front, the arctic weather, and the ferocity and determination of the Russians, who even in retreat took a heavy toll of our men. Todt derided the official optimism about the Russian campaign as willful self-delusion, and predicted that the German people would soon learn the terrible truth.[32]

In November 1941, Todt sent the head of the committee for tank production, Rohland, with a team of armaments industrialists to visit Gen. Heinz Guderian in Orel. There they were shocked to be told that the Eastern Army's equipment and clothing were seriously inadequate for the freezing conditions in the USSR. Tanks and vehicles were freezing up, and soldiers were reduced to wrapping themselves in blankets on the march to try to keep warm.

On Rohland's return, he confronted Todt in a conference on November 28, concluding that the war against Russia could not be won. The following day Rohland and Todt met with Hitler. Rohland repeated his findings in Russia, and adding to that his knowledge of British and American industry, he concluded: "Once the United States entered the conflict, there would be no way of winning the war."[33]

Todt insisted to Hitler that the war could not be won by military means, only politically. Hitler had already discussed this possibility with Goebbels in August 1941, but now—with negotiations already underway with Japan—he had other ideas. The director of the Army's armaments effort, Gen. Friedrich Fromm, was coming to the same conclusion as Todt: peace would have to be made.

From this overall situation, I conclude once again: Dr. Todt—formerly Hitler's "miracle man"—had failed to provide the necessary arms production to win the war in the east after almost two years in office. Something had to be done—fast.

iether Germany would continue to fight, or negotiate, it needed to act from a position of strength, and consequently Todt dedicated himself to rallying Germany's industrialists around the war effort to aid the rebuilding of the Wehrmacht's fighting power. In December 1941, Todt reorganized the regional and national armaments committees into five main committees: ammunition, weapons, tanks, engineering, and general Wehrmacht equipment. This last committee brought together all the many Wehrmacht suppliers who were not involved in the production of armaments. He chose Wilhelm Zangen, head of the RgI, to chair this general committee. He also considered creating a committee for naval procurement. At this time he also formed a new Ministerial advisory Committee manned with representatives from industry and the Luftwaffe.

In January 1942, Todt addressed the leaders of industry at the council of the RgI, and in early February chaired a three-day conference with all the leading armaments producers.

January 1942 also saw Hitler and Goebbels having ongoing discussions about the defeatism in the higher echelons of the regime. Hitler's declaration of war on the United States meant that, "The economic and military forces arrayed against the Third Reich by early 1942 were overwhelming: the world's greatest land army [Russia], fleet [Great Britain], and industrial war-making capacity [the United States]."[34] Despite their loyalty to Hitler, those most closely involved with the German war effort were under no illusions. In particular, "[Ernst] Udet of the Luftwaffe, Fromm of the Army, Thomas of the Wehrmacht High Command, Todt in the Armaments Ministry, Canaris in intelligence, Rohland and his colleagues in the Ruhr, all came to the same conclusion."[35] They all realized that the mobilization of the United States economy could leave Germany in a worse state than in 1918.

On February 7, Todt went to Rastenburg to see Hitler. It was to be his last meeting with the Führer. There are no surviving records of the conversation, but comments from those in Rastenburg at the time suggest it did not go well. Dr. Schmidt has used comments from Rohland to infer some of the discussion:

> Todt … regarded the war with Russia as a national calamity… Todt's desire to build a highway to Russia did not mean that he approved of Hitler's military goals. Despite his NS convictions, Todt would not forgo the right to have his own opinion. He fully realized that Germany—especially after the United States entered the war—ultimately could not keep pace with the arms potential of the enemy…
>
> The crucial thing for him was the fact that Germany had to prepare for a long war, and that from month to month, the prospect of winning such a war grew dimmer and dimmer.
>
> Todt made no bones about his attitude toward the war, whether talking to the Führer or others… After the war, Rohland stated that Todt had repeatedly urged Hitler to end the war politically. Rohland

also presumed that a political solution to the war was the topic of conversation on the evening of February 7, 1942.[36]

According to Dan van der Vat: "Hitler had an irrepressible respect for real experts such as Todt, but he also had an eerie ability to sap the will of even the most independent-minded interlocutors, who therefore avoided a tête-a-tête with him whenever they could … under four eyes, with just two people present, the terms on which Hitler received Todt for the last time on that bitter February evening. Todt, dogged rather than histrionic, armed himself with a paper unfavorably comparing German with Allied industrial capacity in some detail. The unrecorded meeting lasted some six hours, and must have been stormy, because aides heard raised voices through the closed doors."[37]

Adam Tooze surmises: "It is possible that Todt reminded Hitler of their conversation the previous November, and that this provoked an outburst from Hitler, but this is an unsubstantiated claim."[38]

Speer was also at Rastenburg to see Hitler. In an interview in 1971 he recalled seeing Todt after his meeting with the Führer:

As I arrived, Todt emerged from a conference with Hitler, appearing exhausted and depressed. He sat with me, drinking a glass of wine, glumly reticent, and then excused himself for a few hours' sleep. He was flying back to Berlin, he said, and asked me if I'd like to accompany him. I accepted eagerly, glad to avoid the grueling train journey, and agreed to meet him at the airport later.[39]

Speer's meeting, however, went on late into the night, and he decided he could not return to Berlin first thing in the morning. He was awoken the next morning by the telephone, and was informed by Hitler's physician that Todt's plane had crashed, and he had been killed.[40]

Previously unpublished German wartime artwork of a German Navy cruiser in action. Such surface units as these gave way in the main to submarine warfare from 1942. (US Army Combat Art Collection)

CHAPTER FIVE

A FUNERAL IN BERLIN, 1942

"A shattering piece of news reached me"
The Goebbels Diaries, February 9 (probably dictated February 8), 1942

Dr. Todt's flag-bedecked coffin, complete with steel helmet, cushions of medals, and dress sword. The Honor Guard consists of men of his OT. The eagle over the doorway—designed by sculptor Kurt Schmid-Ehmen—was salvaged in 1945 by Col. Edmund W. Hill, and is now displayed at the US Air Force Museum at Dayton, OH. The Russians used Speer's precious red marble to line their rebuilt Berlin subway walls after the war. (Previously unpublished photo, HHA)

On February 9, 1942, Nazi Minister of Propaganda and Public Enlightenment Dr. Paul Josef Goebbels made a somber, but heartfelt and emotional, diary entry for such a normally cynical man:

In the course of the day, a shattering piece of news reached me. Dr. Todt crashed and was killed on leaving the airport of Rastenburg this morning following a visit at General Headquarters. The plane dropped from a height of 400m and exploded on the ground.

The passengers were so badly burned that it was hardly possible to gather up the corpses. This loss is absolutely overwhelming. Todt was one of the really great figures of the National Socialist regime. A product of the Party, he fulfilled a number of historic tasks, the effects of which cannot even be estimated at the moment.

With the genial sparkplug of his personality, he combined an extremely pleasing simplicity of behavior and an objectivity in his approach to his work in so compelling a form that everybody could not but esteem and love him.

The ensuing months will show what we have lost in him. The Führer, too, is hard hit by this loss. We have recently had to endure such heavy personnel losses that one really begins to believe that troubles never come singly.[1]

Perhaps Dr. Goebbels was referring here to the deaths the previous November of Col. Werner Mölders and Col. Gen. Ernst Udet, both leading figures in the German Luftwaffe. Mölders, too, had died in a tragic air crash, and this was given out as well to the public as the cause of Udet's death, although in reality he had committed suicide over the worsening aerial situation of the war.

Later that same day, Goebbels continued his secret diary entry:

I spent the entire day preparing the funeral ceremony for Todt. A solemn State Funeral in Berlin is to honor him before the entire world. The Führer wants personally to come to Berlin to pay him a tribute on behalf of our nation, and thereby to confer the last and highest honors on him.

Throughout the day, I feel numb over this loss. I hardly have time to think! There are so many people in public life who are as superfluous as a goiter. Death does not dare touch them, but when there is someone among them who has the ability to make history, a senseless and cruel fate tears him from our ranks and he leaves a void that simply cannot be filled.[2]

On the next day, Dr. Goebbels added:

Generally speaking, Dr. Todt's fatal crash is regretted throughout the world. Even the English feel they must pay a tribute to his great talent for organization. Of course, there are also a few English voices stupid enough to doubt the official version of the cause of Todt's death. That's the way English gentlemen are: they are nonchalant and polite as long as everything is well with them, but they cast off their masks and reveal themselves as brutal world oppressors the moment one trespasses on their preserves, or a man appears on the scene with whom they must reckon...[3]

Dr. Todt's flag-draped coffin is borne past the Führer's headquarters entourage at the Wolf's Lair as it is loaded from the gun carriage at center for its trip to Berlin for the state funeral of February 11, 1942. Seen at the far right is Nazi Reichsleiter Martin Bormann, a bitter enemy of Dr. Todt's, one of the few he had in the Third Reich. (HHA/Rear Adm. Karl-Jesko von Puttkamer)

Dr. Todt's funeral cortege—draped with the red-black-white Reich War Flag, designed by Speer—arrives in Berlin from the Führer's Wolf's Lair Eastern Front military headquarters, mounted on an artillery caisson. (HHA)

On the 11th, Goebbels wrote:

> The Führer returned to Berlin for the funeral ceremony in honor of Reich Minister Dr. Todt. I went to see him immediately at noon, and noted that the loss of Dr. Todt has shaken him badly... Todt was one of the men closest to the Führer... The Führer told me privately how hard this loss had struck him and how sad he was at the idea of one friend after another gradually leaving our group...
>
> At 3p.m., the state funeral for Dr. Todt was held in the Mosaic Hall of the Reich Chancellery. Every person of prominence in the State, Party, and Wehrmacht attended. One could weep at the thought that so valuable and indispensable a collaborator has been taken from our ranks in such a senseless way. The Führer gave most eloquent expression to this idea in his address.
>
> As he spoke, the Führer was at times so deeply moved that he could hardly continue, but that made all the deeper impression upon those present, and no doubt upon the public as well. We all have the distressing feeling of taking leave of a man who belonged to us as though he were a part of us...[4]

Here are some excerpts from Hitler's eulogy:

> In the sad hour of this ... it is very hard for me to think of a man whose deeds speak louder and more impressively than words can do. When we received the terrible news of this misfortune, to which our dear Master Builder Dr. Todt had fallen victim, many million Germans had the same feeling of emptiness which always occurs when an irreplaceable man is taken from his fellow men...
>
> The work ... this man accomplished with a minimum of assistance. He was without doubt in this field the greatest organizer whom Germany, the German people, has produced up to now. He

managed with the smallest conceivable staff of his own, and without any bureaucracy, to utilize all the agencies and forces which appeared useful ... for the solution of his problems.

Much of what the man has done can be made known to the German people or brought to the amazed attention of the world only after the war. What this man has created is so unique that we all cannot thank him enough for it!

If, however, I spoke just now about the technician and organizer, Fritz Todt, I must also bear in mind the man, who has stood so near to us all. It is not possible to give any better characterization of his personality than in determining that this great director of work never has had an enemy, either in the movement, or among his co-workers.

I myself must especially thank him for the fact that he has never lost or abandoned the ideological heritage of National Socialism, the aims of the movement, in the excess of his responsibilities, but—on the contrary!—has been a co-creator of our world of ideas. And this applies particularly to his attitude toward social problems in life, the man, who himself has directed millions of workers, was not only understanding, but above all in his heart a true Socialist!

There was a time when fate forced him—the greatest construction engineer of all times!—to earn his daily bread as a simple laborer, just as had happened in my own case. Never for a moment was he ashamed of that fact. In later years, it was for him a source of proud and satisfying memories, when he—the greatest construction chief the world has ever known!—had occasion to look at or to show others a photograph of himself depicting him in his sober working attire, working on the road, covered with dust and dirt, or in front of a seething vat of tar.

For this reason, he especially took to his heart his German "road builders," as he called them. It was his continuous desire to improve their social and often so trying living conditions, to replace their former miserable tents with modern barracks and shelters, to take away from the road workers' camps the character of stagnant mass quarters, and especially to create within the laborer the feeling that road building— yes, the entire field of construction—is a field of work of which anyone can always be proud, but it creates documents not only of the highest importance to mankind, but also of the greatest durability.

Before Dr. Todt, the work of the road worker was not regarded very highly. Today, the 10,000 road builders are a proud fraternity fully aware of their great usefulness. In this way, he has accomplished a basic NS educational work, and for this we are today especially indebted to him.

Just as every human progress has had its model, so the "Todt Organization" has created permanent social models, and it was on its way to develop them still further. Gradually, not only a social injustice, but also a human, thoughtless folly, was to be eliminated—and eliminated, indeed, forever.

The new German Order for the Highest Merit with crossed swords and ribbon that was first awarded posthumously by Hitler to Dr. Todt, and affixed to the cushion of medals at his State Funeral in Berlin. There are nine other known awards of the German Order, most of them posthumous. As the Führer noted in his so-called "table talk" to his intimates at FHQ at midday on May 15, 1942, "The State order which he had instituted on the occasion of Dr. Todt's death, and made him the first recipient of, had been created exclusively for the highest service which one man could render to the German Reich." The reverse of the medal featured Hitler's signature in gilt. Dr. Todt's medals, awards, and decorations—in descending order of importance—included the German Order for Highest Merit, the German National Prize for Art and Science, the 1914 Iron Cross First and Second Classes, the 1918 Wound Badge in Black, the Honor Cross for Front Fighters, the German Defense Wall Honor Award of November 23, 1939; the Golden Party Badge, the Golden Hitler Youth Honor Cross with Oak Leaves, the Service Badge of the Nazi Party in Silver and Bronze, and the German Olympic Honor Cross First Class, as well as other, foreign awards. (HHA)

Sieg Heil! (Hail Victory!) The Führer gives the Nazi Party salute to the funereal bier and coffin of the deceased Dr. Todt amidst an Honor Guard of OT men and bouquets of flowers. The Führer had just laid his own massive floral wreath at the foot of his late colleague's coffin at one end of the Mosaic Hall in the Reich Chancellery in Berlin. (HHA)

Thus, whether this man had dealings with a working man, a minister, or a general, he always remained the same, an equally confident leader and solicitous friend of all decent national comrades. It was no wonder that this man—who so loved his people!—was passionately attached to his family, his wife, and his children.

The creator of the greatest technical enterprises spent every free hour, whenever he could, among the great creations of nature, in the little house beside the lake, in the midst of his beloved Bavarian peasants.

When under the fire of enemy guns the West Wall was completed, while in Poland the columns of the Todt Organization for the first time joined our advancing armies and gave them assured supply lines, I had it in my mind to award him the Knight's Cross, as one of the leading

February 11, 1942, 3p.m.: The official mourning party for Dr. Todt in the Mosaic Hall. Sitting in the first row, right to left, are Führer and Reich Chancellor Adolf Hitler, Frau Todt, Reich Marshal Hermann Göring, Todt's daughter, Ilsebill; Todt's successor as Minister of Armaments and Munitions, architect Dr. Albert Speer. In the second row, right to left, are Schaub, Goebbels, Adm. Erich Raeder, Keitel, and Col. Gen. Erhard Milch. (Previously unpublished photo, HGA)

Reich Marshal Göring (left), who coveted Dr. Todt's Ministry, salutes with his ornate, bejeweled Reich Marshal's baton as the flag-draped coffin is borne out of the Reich Chancellery's Mosaic Hall. To the left of Göring can be seen the head of Dr. Goebbels, who so lamented Dr. Todt's demise in his secret diary entries for February 8–11, 1942. Note also, Hitler's now empty chair: the German Head of State had already left, leaving the final salute to Göring, his own designated successor as Führer. (Previously unpublished photo HGA)

creators of the German resistance and the German will for self-expression in the war. However, I changed my mind.

Because this distinction—famous though it is—could never have done justice to the importance of this unique man, I had already made the decision some time previously to establish such a decoration, which, founded on the principles of our movement, is to honor in several classes the most valuable services that a German can perform for his people.

After the conclusion of the campaign against France, I said to Dr. Todt that I proposed for him some day, as God wills, the recognition of his unique service, and that he would be the first to whom I would award the highest class of the order. In his modesty at that time, he did not want to know anything about it.

In field gray wartime SS uniform, Gen. Hans Baur (right) laughs with an unknown Luftwaffe officer at left at Hitler's Wolf's Lair FHQ in 1941. The man at center is the Führer's Luftwaffe liaison officer during 1937–45, who later performed the same role for Dr. Speer in a dual function, Col. Nikolaus von Below. It was he who first heard Hitler say that only Professor Speer would be able to replace the dead Dr. Todt on the morning of February 8, 1942, following the fatal air crash. (Previously unpublished HHA photo)

A previously unpublished photograph of Luftwaffe Col. Gen. Ernst Udet (right) with his chief, Air Force Field Marshal Göring before the war. Göring brought Udet into the Luftwaffe in 1933. Udet was out of his depth as head of Luftwaffe technical development, a fact overlooked by Göring, but fully realized by his chief rival within the Air Force, later Field Marshal Erhard Milch. He also favored single-engined Ju 87 Stuka dive-bombers as opposed to four-engined long-range "Ural bombers" that Nazi Germany later desperately needed to win the war in the east and bomb New York, but, alas, never had. That said, however, to his overall strategic credit, Col. Gen. Udet was the first among the top Nazi leadership cadre to believe—in the autumn of 1941—that the Third Reich had already lost the air war. (HHA)

As now the National Order of Art and Science was awarded first to Dr. Todt—no, to the deceased Professor Troost—so now, today, I confer for the first time in the name of the German people and its National Socialist movement, the new order on our dear and unforgettable Party comrade, Dr. Todt, the General Inspector of our Roads and builder of the West Wall, the organizer of armaments and munitions in the greatest battle of our people for their freedom and their future.

I can add only a few words for myself. I have lost in this man one of my most faithful co-workers and friends. I regard his death as a contribution to the National Socialist movement, to the fight for freedom of our people.[5]

In his diary, Dr. Goebbels concurred:

Todt certainly deserved this. If *anybody* had a right to be awarded posthumously the highest honors that the Reich can confer, it was *he!*

HITLER'S FÜHRER HEADQUARTERS 1939–45

The afternoon that World War II began, Hitler left Berlin aboard his Special Train *Amerika* (later codenamed *Brandenburg*), his first field *Führerhauptquartiere* (Führer Headquarters, or FHQ) of the war. He would use it briefly again for such a purpose during the Balkan campaign of May–June 1941. During the war, Hitler would deploy to a series of far-flung, top-secret temporary field FHQs across Nazi-occupied Europe. The first series—during the period 1940–41—were either already existing structures, or in rather small, quaint, log cabins and huts in Germany, Belgium, and France.

Initially Castle Ziegenberg was outfitted for Hitler as his first Western Front FHQ in 1940, but the Führer rejected it, being reminiscent of the plush settings favored by Kaiser Wilhelm II during World War I. During the Battle of the Bulge campaign of December 1944–January 1945, German Army Field Marshal Gerd von Rundstedt used it as his field headquarters instead, while Hitler settled nearby.

The second series—in both German East Prussia (formerly Poland) and Nazi-occupied Ukraine during 1941–44—started out in roughly the same fashion, but on a far larger scale than before. In early 1944, the most famous of all the FHQs—the Wolfsschanze (Wolf's Lair)—underwent a second and final, massive concrete and steel facelift, in which all of the original buildings were vastly enlarged yet again, and fully encased, to protect against enemy aerial bombardment.

In the main, the bombardment never happened, and nor did what Hitler feared most of all: an Allied paradrop to either murder or kidnap him.

His final two FHQs receded once again to the Western Front: at Adlerhorst (the Eagle's Nest) near Castle Ziegenberg, and, lastly the connected New German Reich Chancellery in Berlin and its underground bunker complex, the Führer Bunker.

All of the field FHQs, however, were designed and built by troops of the OT, first under Dr. Todt, and then Professor Speer. Actual site selections were made by a team of his military adjutants, principally his OKW aide, Maj. Gen. Rudolf Schmundt; Army aide-de-camp Maj. Gerhard Engel, and his Luftwaffe liaison officer, Col. Nikolaus von Below, all working in concert with Drs. Todt and Speer. Indeed, in 1944, one of them—von Below—pulled double duty in his adjutant's role, retaining that function under his Führer, but also taking on the role of liaison officer at FHQ for Professor Speer as well.

According to Leon Krier, "The *Baustab Speer* (Building Staff Speer) in the OT was responsible for Hitler's headquarters on the Eastern and Western Fronts. A large number of buildings—integrating old castles and manor houses—were camouflaged as spontaneous ensembles and built in the local vernacular. In case of aerial attack, identical technical and domestic equipment was available in an underground bunker system."[6]

A trio of top-ranking SS men at one of Hitler's 1940 Western campaign FHQs during the first year of the war. From left to right: Karl "Wolffshen" Wolff, the Reichsführer-SS Heinrich Himmler, and Reich Chancellery chief Dr. Hans Heinrich Lammers. Because he engineered the surrender of northern Italy to the Allies on April 29, 1945—a week before the war ended—Wolff was spared the hangman's noose, and died of natural causes in 1984. His memoirs have not yet been translated into English. (HHA.)

The ceremony was sad and sorrowful. When the remains of Todt were carried out of the Reich Chancellery, it seemed to all of us that a brother was leaving us.

On this day, I was not really in the mood to attend to all sorts of petty details. The Führer, too, retired. I occupied my mind with a number of problems that will assume importance only in the distant future. Work is always the best antidote to attacks on the soul and spirit. I passed the evening in a melancholy mood...[7]

Among the top men sitting in the great red marble Mosaic Hall—designed and built by himself, no less—was one of Hitler's personal and favored architects, the General Building Inspector for Berlin, Professor Dr. Albert Speer, the successor-designate of the man whose funeral he had just witnessed. At 36, Dr. Todt's successor was the youngest cabinet minister in the Third Reich, and now, by all accounts, had some very large shoes to fill.

Accident or Nazi Murder Plot?

With Dr. Todt's funeral over, the question turned to *how* the fatal crash had occurred. Rumors had been rife since Todt's death had been announced to the public in a brief statement: "On Sunday, February 8, in the fulfillment of his duties as a soldier, Reich Minister Dr. Todt died in an accident while carrying out his military assignments."[8]

The public was shocked, and the style of the announcement led to criticism. Rumors soon began to spread, and it was generally concluded that the accident had been caused by sabotage, or arranged by the enemy.[9] The Führer and those closest to him also wanted to know what had happened, as Speer recalled:

On Hitler's orders, the Reich Air Ministry tried to determine whether sabotage might have been responsible for the plane crash.

The investigation established the fact that the plane had exploded, with a sharp flame darting straight upward, some 65ft above the ground. The report of the commission, which because of its importance was headed by an Air Force lieutenant general, nevertheless concluded with the curious statement: "The possibility of sabotage is ruled out. Further measures are therefore neither requisite nor intended."

Incidentally, not long before his death, Dr. Todt had deposited a sizable sum of money in a safe, earmarked for his personal secretary of many years' service. He had remarked that he was doing this in case something should happen to him.[10]

The question, naturally, immediately arises: did Todt take this action following a premonition that he might be murdered, either by the Allies, or by some jealous competitor from within the Nazi hierarchy? If the latter, who would the likely suspects include? The list could be

Colonel General Udet shot himself, taking to his grave the fact that the Luftwaffe was already a strategic failure. A playboy, womanizer, gambler, and heavy drinker by nature, Udet cracked under the strain of being a desk officer within Göring's Air Force High Command. Reich Marshal Göring walked behind Udet's caisson-borne casket as chief mourner, followed by Luftwaffe combat aces; Knight's Cross holders made up the honorary escort and pallbearers. The Reich Marshal delivered the funeral eulogy: "I can only say that I have lost my best friend." (Previously unpublished photograph, Hermann Göring Albums, LC, Washington, DC)

endless. Himmler, Göring, Martin Bormann, and even Speer himself, were all jealous of his power, while at the same time increasing their own and competing with each other.

There was also Hitler himself, who might have feared Dr. Todt not only as a future critic of the regime (as, indeed, Mussolini had feared Marshal of the Air Force Italo Balbo before *his* death in a June 1940 air crash), but also as a possible better-equipped successor to him as Führer (as the Duce also saw in Balbo).

There was, even then, an anti-Hitler movement in the Reich and among the various far-flung military commands that had been conspiring to overthrow the Nazi regime since 1933 and, as we shall see, there was Dr. Speer, too.

The possibility of a suicide triggered by depression—as had been the case with Col. Gen. Ernst Udet in November 1941, and would be with Luftwaffe Chief of the General Staff Col. Gen. Hans Jeschonneck in August 1943—cannot be ruled out entirely, either.

In a postscript, Professor Speer continued his own narrative on the crash details:

> The plane executed a normal takeoff, but while still within sight of the airport, the pilot made a rapid turn, which suggested that he was trying for an emergency landing. As he was coming down, he steered for the landing strip without taking time to head into the wind. The accident occurred near the airport, and at a low altitude.
>
> The plane was a Heinkel 111, converted for passenger flight; it had been lent to Dr. Todt by his friend, Field Marshal [Hugo] Sperrle, since Todt's own plane was undergoing repairs. Hitler reasoned that this Heinkel, like all the courier planes that were used at the front, had a self-destruct mechanism on board. It could be activated by pulling a handle located between the pilot's and the co-pilot's seats, whereupon the plane would explode in a few minutes.[11]

The final report of the military tribunal, dated March 8, 1943 and signed by the commanding general and the commander of Air District 1, Königsberg, stated:

> Approximately 2,300ft from the airport and the end of the runway, the pilot apparently throttled down, then opened the throttle again two or three seconds later.
>
> At that moment, a long flame shot up vertically from the front of the plane, apparently caused by an explosion. The aircraft fell at once from an altitude of approximately 65ft, pivoting around its right wing and hitting the ground almost perpendicularly, facing directly away from its flight direction. It caught fire at once, and a series of explosions totally demolished it.[12]

It should be noted that this accident involved a Luftwaffe plane exchanged for another Luftwaffe plane by a Luftwaffe field marshal, and was later investigated by a Luftwaffe commission, all under the overall command of the commander in chief of the Luftwaffe, Reich Marshal Hermann Göring, who was Dr. Todt's rival for power. These are facts which may or may not have direct bearing on the question of whether or not Dr. Todt was murdered in an internal Nazi plot, or perhaps on Göring's own orders. As we shall see, Göring *was* a significant player in the drama that unfolded immediately *after* the crash.

In 1970, Professor Speer himself raised the question of foul play in Dr. Todt's sudden demise:

> At first, Hitler seemed to treat Todt's death with the stoic calm of a man who must reckon with such incidents as part of the general picture. Without citing any evidence, he expressed the suspicion—during the first few days—that foul play might have been involved, and that he was going to have the Secret Service look into the matter. This view, however, soon gave way to an irritable and often distinctly nervous reaction whenever the subject was mentioned in his presence. In such cases, Hitler might declare sharply: "I want to hear no more about that. I forbid further discussion on the subject." Sometimes he would add, "You know that this loss still affects me too deeply for me to want to talk about it."[13]

What are we to make of this passage? Was Hitler simply playacting for the benefit of his entourage, and/or did he know more than he cared to reveal? If one posits the SS as a possible rival—under its chief, Reichsführer-SS Himmler—and therefore a likely candidate for the murderer, then having "the Secret Service look into" the matter is tantamount to having the criminal investigate himself!

The possibility of the SS as the culprit simply cannot be overlooked, especially as Reichsführer-SS Himmler was in direct rivalry for Dr. Todt's own power base, as detailed in Professor Speer's last book, *Infiltration*.

The man in the middle was the often unhappy role played by Luftwaffe Col. Nikolaus von Below, seen here (center) in the Führer's ornate, grandiose office in the NRC in Berlin as his nominal chief, Air Force C-in-C Field Marshal Göring (left), presents fiftieth birthday greetings to Reich Chancellor Hitler on April 20, 1939 in this previously unpublished photograph. Colonel von Below answered to three major Nazi leaders by war's end: Hitler, Göring, and Dr. Speer, serving both the Führer and his Armaments Minister as FHQ Air Force liaison officer, as well as still reporting to Reich Marshal Göring—most confusing! (HGA)

As for the Heinkel 111's self-destruct mechanism, Dr. Schmidt stated:

> It is not out of the question, according to Speer, that the pilot inadvertently pulled the self-destruct lever himself. In Todt's plane—which had no such device—the identical switch had an entirely different function, but in Sperrle's combat plane, the mistake was fatal... It took three minutes for the mechanism to work. The pilot did not have enough time to return to the airport and land before the kilogram of explosives blew up. Postwar investigators who attempted to reconstruct the disaster by using an airplane carrying a time bomb only fed new rumors. The SS and even Hitler himself were suspected of trying to assassinate Todt.[14]

Schmidt however thought that Hitler was unlikely to have killed Todt, for despite his speaking out against the Führer, he was a talented man whom Hitler had esteemed and respected for many years. However, "Rumors about the SS responsibility for Todt's death persist even today. A top leader who felt it made no sense to prolong the war might have been too uncomfortable a prophet for the SS. Perhaps in that black-uniformed élite, one might find the wire-pullers of a 'deliberate' accident. Be that as it may, the death of Fritz Todt will probably always remain one of the unsolved enigmas of history."[15]

Of course, Speer had an agenda in his conclusions on the death of Todt. His memoirs, *Inside the Third Reich*, and later works, set out his own view of events, and were published after most of the protagonists were dead. Works such as Schmidt's biography set out expressly to debunk the architect's own view of history. Professor Speer's account of the Dr. Todt airplane crash did not appear until 1969 in German, and in translation in 1970, after he had served two decades in Spandau Prison.

The memoirs of SS Lt. Gen. Hans Baur, Hitler's own personal chief pilot during 1932–45, were not researched and written until he was released in 1955, after his own decade-long imprisonment

In newly occupied Austria on March 26, 1938, Luftwaffe Gen. Hugo Sperrle (left) greets his commander in chief, Field Marshal Göring. On the morning of Dr. Todt's death, the Reich Armaments Minister flew out of Rangsdorf Airfield in Sperrle's lent aircraft, not his own, perhaps contributing to his sudden death. Austrian Luftwaffe Gen. Alexander Löhr salutes between them. (Previously unpublished photo, HGA)

within the Soviet Union. These were first published in English in 1958 under the title *Hitler's Pilot*.[16] Although Dr. Speer could have accessed Baur's memoirs, there is no mention at all of the work in his book. However, Dr. Schmidt *does* present General Baur's view, but only summarized in one paragraph.

Speer may have chosen not to mention Baur for a variety of reasons. First, according to the crusty ex-World War I pilot, the two men disliked each other by 1945. Second, according to Professor Speer's publisher, Macmillan Company in New York, the "original manuscript notes ... were written during his first years in Spandau Prison, and are the basis for these memoirs."[17] In which case, Speer would not have known of Baur's version of events when he was writing his memoirs.

This author prefers to present General Baur's viewpoint in full, as a fitting conclusion to this incident:

Reichsminister Dr. Todt, a welcome guest, was well liked by all. He was greatly admired because, in spite of his great talents, he remained modest. On the day before his death, he had delivered a long report to Hitler. I spent the late evening hours with him. He told me that he wanted to return to Berlin early the next morning, and that he might take Speer, who was also there visiting Hitler, with him.

On the next morning, I drove to the airport at about 8:30, as usual. From far off, I saw a cloud of black smoke in the vicinity of the airfield, and shortly, about 50m from the field, came upon a burning Heinkel. I learned that it was Dr. Todt's plane.

The Heinkel had just taken on 3,400l of gasoline, so the fire would continue to rage for some time. I sent for long poles to lift the bodies out of the fire before they were burned to ashes. Several fire extinguishers directed their streams at the fire, reducing the heat somewhat. We succeeded, using the poles, to pull out the corpse.

Dr. Todt lay head downward, one shoulder touching the earth, and one part of a shoulder that definitely identified the body as Dr. Todt still recognizable. I discovered that Speer had not flown with him. Besides Todt and the three-man crew, a couple of vacationers had boarded the plane. They all burned to death.

Naturally, I immediately took reports from the witnesses in an effort to determine the cause of the crash. The weather conditions on that day were very bad: a heavy storm, clouds to the height of 2,300m, and occasional snowfall.

Major Mölders, Deutschlands erfolgreicher Jagdflieger,

French General Charles de Gaulle noted that following World War II, "The graveyards of the world are filled with indispensable men." So it was, too, with the Third Reich's most celebrated ace at the time of his death in an air crash in the fall of 1941. He was, ironically, on his way to be a pallbearer at Colonel General Udet's funeral in Berlin. This was Göring's first General of the Fighters, Col. Werner Mölders, seen here being lionized in the Nazi press on the cover of the *German Illustrated* magazine on October 8, 1940 celebrating his 40th aerial kill. Colonel Mölders was credited with starting the fighter pilot's "finger four" formation of aerial combat during his prewar service in the Spanish Civil War as part of the famed Kondor Legion, the same formation still used today by air forces the world over. (HHA)

The flight policeman had just made a note of the exact takeoff time. Three minutes later, he noticed the plane returning. That was unusual, and what surprised the man even more was that the pilot was already approaching for a landing with the landing gear down.

The thought ran through his head that, with such a strong tailwind, a landing had little chance of success because the remaining runway was too short. Then suddenly, when the Heinkel was only 100m from the border of the field, a blue jet of flame shot out of the rear of the plane. That jet of flame was also seen by people on a neighboring estate.

On the Heinkel, the fuel was carried in the wings, so a gasoline explosion could not have been the cause. The witnesses unanimously agreed that the flame had belched from the plane's rear section.

We found it hard to assume that it was sabotage. Our personnel at the airfield were dependable, and furthermore, Dr. Todt was extremely well liked. It could hardly have been an outsider because the guard was heavy and most trustworthy, so we were faced with a riddle. A commission from the German Ministry for Air Travel investigated all the fragments of the wreckage, and they, likewise, arrived at no conclusive results.

The cause was still unknown. Finally, the flight records of the plane were examined. Every airplane has its papers in which

Luftwaffe Commander in Chief Reich Marshal Göring holds aloft his jewel-encrusted baton of office in high salute as the gun carriage bearing the coffin of Mölders passes him during the State Funeral in Berlin on November 30, 1941. Coming as it did on the very day of Colonel General Udet's funeral, Colonel Mölders' tragic death robbed the Reich of its greatest aerial ace, and caused renewed gossip as to the accident's cause. The rumor mill began anew at the sudden demise of Dr. Todt. (Previously unpublished photo, HGA)

everything about the plane is listed: changes of engines, repairs and checks. The following picture emerged: Dr. Todt's plane had gone in for a major checkup in which the plane was disassembled, and parts individually checked, and damaged or worn parts were replaced.

Another Heinkel from an Air Force unit had been placed at Dr. Todt's disposal for the 14 days required by this process.

This replacement unit had been at the front, and, like all combat area aircraft, it carried the so-called "destroyer," a small box containing a kilogram of dynamite activated by a pull string with a tiny loop. The "destroyer" was located under the pilot's seat. In emergencies, the airplane could be blown to smithereens. A built-in time fuse gave a three-minute delay. This "destroyer" probably sealed the minister's fate.

Dr. Todt usually sat in the cockpit next to the pilot in the place normally occupied by the flight engineer. Shortly before takeoff, he walked through the little cabin to the pilot's seat. The passage was very narrow, and Dr. Todt—wearing a fur suit—had to squeeze through the opening. He sat and waited until the flight engineer, who also served as radio operator, had raised the landing gear and could vacate the other seat.

In squeezing his way from the passage to the seat, it is possible that the loop on the pull cord caught on one side of the buttons of his boot, thereby setting off the timer and the detonator. The burning of the delay fuse would have produced smoke in the cabin, setting off an automatic alarm. This would have initiated a search for the cause of the smoke, which the crew probably found rather quickly.

The airplane had probably been in the air about two minutes when the problem was discovered. Then began a race with death in which the outcome depended upon a few critical seconds. The landing gear was lowered. There was no time to lose. The landing had to be attempted in spite of the heavy tailwind.

The airplane only reached the boundary of the airfield when the race was lost by seconds. The "destroyer" exploded.

The plane flipped over at an altitude of 30m, crashed, and burned completely. Hitler, to whom I had to give detailed reports on the event, felt the loss very deeply.[18]

I leave the final conclusions up to the reader. For my own part, however, I am inclined to believe that General Baur was correct in his depiction of events, and Dr. Todt's death was, therefore, no Nazi assassination plot, but an accident, pure and simple. To my mind, then, there is only one final question: what happened to the German Air Ministry investigative report?

For instance, but five months later, in the June 1942 SD investigation into the assassination of SS Gen. Reinhard Heydrich in Prague, maps were drawn, sketches made, and photographs taken from every conceivable angle. Surely the same was done in this case with so important a personality as Dr. Todt?

If the report was filed with the German Air Ministry in Berlin, it may have been destroyed by Allied bombing later in the war, but a copy may well have been kept elsewhere. At the end of the war, most of Nazi Germany's state papers, records, and archives fell into the hands of the US Army intact, so perhaps this investigation report survived and will yet reemerge. In its absence, this author feels that Dr. Fritz Todt was *not* the victim of an assassination plot from any quarter, but rather that the death of the road builder on February 8, 1942 was a tragic accident, and one that led to Dr. Speer's rise, subsequent fall, imprisonment, and enduring fame.

Enter Speer

With Todt's death, Speer's hour had come. Speer often told others how easily he could have shared Todt's fate, as he had been scheduled to be on the very same flight with Dr. Todt, but had decided instead to sleep in that morning following his own late-night meeting with Hitler.[19] Decades later, Speer depicted death as having enabled his rise to power:

How often my career was affected by death, for I would never have become Hitler's architect but for the death of Troost, whose designs he so admired, and certainly he would never have appointed me Minister of Armaments if my predecessor, Todt, had not been killed in an airplane crash.[20]

Speer stated that he was awakened by a telephone call from Hitler's surgeon—SS Dr. Karl Brandt—informing him of the tragedy. He describes, from his own perspective, what happened that morning at the Führer's headquarters:

At the breakfast table in the Führer's headquarters there was lively discussion of who could possibly be considered for Dr. Todt's successor.

Everyone agreed that he was irreplaceable, for he had held the positions of three ministers. Thus, he had been the supreme head of all roadbuilding operations, in charge of navigable waterways, and improvements on them, as well as of all power plants. In addition, as Hitler's direct envoy, he was Minister of Armaments and Munitions.

Within the framework of Göring's Four Year Plan, he headed the construction industry, and he also created the Todt Organization...

During these first few hours, I had already realized that an important portion of Todt's widely ranging tasks would surely fall to me, for as early as the spring of 1939, on one of his inspection tours of the West Wall, Hitler had remarked that if anything should happen to Todt, I would be the man to carry out his construction assignments.

Later, in the summer of 1940, Hitler had received me officially in the Chancellery office to inform me that Todt was overburdened. He had therefore decided, he said, to put me in charge of all construction,

The top politico-military leadership of the Nazi Third Reich sing the German national anthem—*Germany Over All*—at this wartime Party rally in the Berlin Sports Palace. Seen here from left are Führer and Reich Chancellor Hitler, Reich Marshal Göring, OKW chief Army Field Marshal Wilhelm Keitel, Luftwaffe Field Marshal Erhard Milch, and Reichsführer-SS Heinrich Himmler, in his role as head of the Waffen (Armed) SS. Note the Reich War Flag at left, designed by Dr. Speer. (Previously unpublished photograph from the Heinrich Hoffmann Albums)

including the fortifications along the Atlantic. At the time, I had been able to convince Hitler that it would be better if construction and armaments remained in one hand, since they were closely linked.

Hitler had not referred to the matter again, and I had not spoken to anyone about it. The arrangement would not only have offended Todt, but surely would have diminished his prestige.[21]

Of course, in all of these matters of discussion "under four eyes," as Hitler described them, we have only Dr. Speer's word for this. Continuing, Speer described how later the same day he met with Hitler:

I was, therefore, prepared for some such assignment when I was summoned to Hitler as the first caller of the day at the usual late hour, around one o'clock in the afternoon.

Even the face of Chief Adjutant [Julius] Schaub expressed the importance of the occasion. In contrast to the night before, Hitler received me officially as Führer of the Reich.

Standing, earnest and formal, he received my condolences, replied very briefly, then said without more ado: "Herr Speer, I appoint you the successor to Minister Todt in all his capacities." I was thunderstruck. He was already shaking hands with me and on the point of dismissing me, but I thought he had expressed himself imprecisely and therefore replied that I would try my best to be an adequate replacement for Dr. Todt in his construction assignments.

"No, in all his capacities, including that of Minister of Armaments," Hitler corrected me. "But I don't know anything about..." I protested. "I have confidence in you. I know you will manage it," Hitler cut me off. "Besides, I have no one else. Get in touch with the Ministry at once and take over!" "Then, mein Führer, you must put that as a command, for I cannot vouch for my ability to master this assignment."

Tersely, Hitler issued the command. I received it in silence. Without a personal word—such as had been the usual thing between us—Hitler turned to other business. I took my leave, having experienced a first sample of our new relationship. Hitherto, Hitler had displayed a kind of fellowship toward me as an architect. Now, a new phase was perceptibly beginning. From the first moment on, he was establishing the aloofness of an official relationship to a minister who was his subordinate.

As I turned for the door, Schaub entered. "The Reich Marshal is here, and urgently wishes to speak to you, mein Führer. He has no appointment." Hitler looked sulky and displeased. "Send him in." He turned to me. "Stay here a moment longer."

Göring bustled in, and after a few words of condolence, stated his mind: "Best if I take over Dr. Todt's assignments within the framework of the Four Year Plan. This would avoid the frictions and difficulties we had in the past as a result of overlapping responsibilities."

Göring had presumably come in his special train from his hunting lodge at Rominten, about 60 miles from Hitler's headquarters. Since the accident had taken place at 9.30a.m., he must have wasted no time at all.

Hitler ignored Göring's proposal. "I have already appointed Todt's successor. Reich Minister Speer here has assumed all of Dr. Todt's offices as of this moment." The statement was so unequivocal that it excluded all possible argument. Göring seemed stunned and alarmed, but within a few seconds recovered his composure.

Coldly and ill-humoredly, he made no comment on Hitler's announcement. Instead, he said: "I hope you will understand, mein Führer, if I do not attend Dr. Todt's funeral. You know what battles I had with him! It would hardly do for me to be present."

I no longer remember precisely what Hitler replied since all this washing of dirty linen was naturally somewhat of a shock to me at this early moment in my new ministerial career, but I recall that Göring finally consented to come to the funeral, so that his disagreement with Todt would not become public knowledge.

Given the importance assigned to such ceremonies by the system, it would have caused quite a stir if the second man in the state was absent from a formal act of state in honor of a dead cabinet minister.

There can be no doubt that Göring had tried to win his point by a surprise assault. I even surmised that Hitler had expected such a maneuver, and that this was the reason for the speed of my appointment. As Minister of Armaments, Dr. Todt could carry out his assignment from Hitler only by issuing direct orders to industry.

Göring, on the other hand, as commissioner of the Four Year Plan, felt responsible for running the entire war economy. He and his apparatus were therefore pitted against Todt's activities...

This is the Badge of Honor for the Dr Fritz Todt Prize, instituted on February 8, 1944, the second anniversary of his death, and given for inventions which furthered the war effort. The Dr. Fritz Todt Prize Badge of Honor was awarded in a trio of classes, as signified in gilt, silver, and black steel. Note the honoree's name below the Nazi eagle, as well as the rotating swastika within the cogwheel logo of Dr. Robert Ley's DAF NS labor organization. Stated Littlejohn, Dodkins, and Bender:"The badge is in three classes… The class awarded depended on the importance of the invention, and the degree of the enterprise shown. The gilt badge could be awarded only on the recommendation of Dr. Ley and the Chief of the Technological Department of the Nazi Party. The badge, which was awarded twice annually on September 4th [the date of Dr. Todt's birth], and February 8th [the anniversary of his death], was worn on the left breast pocket." (Courtesy of George Peterson, National Capital Historic Sales, Springfield, VA)

In the middle of January 1942, about two weeks before his death, Todt had taken part in a conference on production matters. In the course of it, Göring had so berated him that Todt informed Funk on the same afternoon that he would have to quit.

On such occasions, it worked to Todt's disadvantage that he wore the uniform of a brigadier general of the Air Force. This meant that, in spite of his ministerial office, he ranked as Göring's subordinate in the military hierarchy.

After this little episode, one thing was clear to me: Göring would not be my ally, but Hitler seemed prepared to back me if I should encounter difficulties with the Reich Marshal.

One can only wonder at the recklessness and the frivolity with which Hitler appointed me to one of the three or four ministries on which the existence of his state depended.

I was a complete outsider to the Army, to the Party, and to industry. Never in my life had I had anything to do with military weapons, for I had never been a soldier, and up to the time of my appointment, had never even used a rifle as a hunter. To be sure, it was in keeping with Hitler's dilettantism that he preferred to choose a non-specialist as his associate.

After all, he had already appointed a wine salesman as his Foreign Minister [Joachim von Ribbentrop], his Party philosopher as his Minister for Eastern Affairs [Alfred Rosenberg], an erstwhile fighter pilot [Göring] as overseer of the entire economy. Now, he was picking an architect of all people to be his Minister of Armaments.

Undoubtedly, Hitler preferred to fill positions with laymen. All his life he respected but distrusted professionals such as, for example, Dr. Hjalmar Schacht.[22]

Minister Speer's account of his appointment is very likely a fantasy concocted after the fact, designed to present its author as a neophyte to the real world of Nazi politics. Gitta Sereny noted that his description of the events surrounding Todt's death changed in each version of his memoirs. "Speer agonized over how to present the story," Sereny asserted.[23]

My conclusion is that he and Hitler had been discussing the problems of the Armaments Ministry since at least the previous summer, after *Barbarossa* began to show signs of being stalled. At once both cautious and impulsive, the Führer had kept Professor Speer in mind, ready for the right moment to present itself.

Speer's presence at FHQ at that time, I conclude, was no accident, but was for a shift of command that might very well already have happened the night *before* Dr. Todt's fatal flight. After all, the previous December, Hitler had duly shaken up the German Army, taking over its Supreme Command himself by firing von Brauchitsch, as well as relieving Field Marshal Gerd von Rundstedt and Col. Gen. Heinz Guderian of their commands.

It might very well have transpired that Speer rushed to Hitler, and even *asked* for the appointment, rather than being summoned, as he related in his memoirs. We shall never know, as only two men were present at the historic encounter, and now both are dead.

One of the first writers to question Dr. Speer's post-Spandau version of events was Dr. Matthias Schmidt in his 1982 biography *Albert Speer: The End of a Myth*, the publication of which Speer tried to block.[24] Schmidt discusses the speed with which Hitler appointed Speer:

Adolf Hitler instituted "The German National Prize for Art and Science" in 1937 after forbidding German citizens from accepting the Nobel Prize. At the second award ceremony, on September 1938, Todt was among the recipients. The awardees, facing away from the camera, are (left to right) Todt, Volkswagen designer Dr. Ferdinand Porsche the Elder, aircraft designers Willy Messerschmitt and Ernst Heinkel. As well as the prize, the recipients got a cash award of 100,000 Reich Marks. (HHA/Peterson)

> When Hitler was informed of the plane crash around 9a.m. that Sunday, his Air Force adjutant, [Col. Nikolaus] von Below, happened to be present. Hitler "reacted … almost imperceptibly. All he said, after a few minutes, was that Professor Speer was the only man who could replace Todt…"
>
> Speer's appointment as Todt's successor has been widely discussed by scholars. The British historian Alan Milward set February 18 as the date of the appointment, and revisionist historian David Irving doubted that Speer was named successor in all of Todt's functions as early as February 8.
>
> Finally, the question remains: *was* Hitler's decision based on sudden inspiration or on his characteristic "dilettantism" … or had he made up his mind after long reflection?

Dr. Speer—who allegedly hated giving speeches more than two sentences long, as a sort of German Calvin Coolidge—was "forced" to address his new ministry staff on February 14, six days after the fatal crash. He chose to do so in the courtyard of the ministry building on Pariser Plaza not far from the Berlin NRC in a driving snowstorm, the site of a recent speech to the staff by Dr. Todt himself, on his 50th birthday.

Professor Speer stated: "At the zenith of his tremendous labors, Party Comrade Dr. Todt was carried off. His work will remain indelibly imprinted for all time in the Book of History… Great as our grief for the dead man may be, the Führer's grief is even greater. With our tireless work, we will all make it easier for him to get over this pain. *Sieg Heil! Sieg Heil! Sieg Heil!*"[25]

Far from modestly turning away from such a powerful post within the very highest strata of the Nazi hierarchy, Dr. Schmidt concluded, Professor Speer actively *sought* it, and in this assessment, I entirely concur. *Did* he plot to take over Dr. Todt's job by *killing* him? Answers elude us conclusively on *all* these questions, as they probably always will.

Dr. Schmidt's own conclusion is that, "Perhaps on that fateful February 8, 1942, Speer hesitated briefly before laying aside his architectural ambitions for his new office. However, the facts show that, once in office, he unhesitatingly grabbed the full wealth of power that his patron Hitler was willing to give him. In his inconspicuous, but coldly grasping way, Speer swiftly managed to fortify the vast authority he gained as weapons-maker of the Third Reich."[26] This author concurs with the last assertion.

CHAPTER SIX

ENTER DR. ALBERT SPEER

"I was swept away!"
Albert Speer, recalling his fascination with Hitler

A wartime portrait of German Minister of Armaments and War Production Dr. Albert Speer wearing the uniform of the Organization Todt that he adopted from 1942. "He was a great organizer" said his former architectural aide Willi Schelkes. (Photo from the Heinrich Hoffmann Albums, US National Archives, College Park, MD)

Having joined the rising Nazi Party on March 1, 1931 as Member #474,481, Dr. Speer also joined the hooligan SA Stormtroopers on the very same date, then left the brown-shirted thugs for the more élite SS Motorized Division in the autumn of 1932, thus making him a subordinate of first SA Staff Chief Ernst Röhm, and then the even more unsavory Heinrich Himmler, chief factotum of the Holocaust against the Jews, gypsies, and others whom the Nazis considered racial inferiors. In his post-Spandau books, Speer went to great lengths to conceal his early membership in this most detested of all Nazi organizations. How could a man like Dr. Speer serve a man like Hitler? It is a question that will most likely never be answered.

Speer was born Berthold Konrad Hermann Albert Speer on March 19, 1905 in Mannheim, Baden, Germany. He was a child of affluence, growing up in industrial Mannheim—a Daimler-Benz Company town. His family boasted a cook, and servants that included a maid, chambermaid, butler, and chauffeur, and a French nanny.

Speer, Sr. was "a neo-Renaissance architectural builder," as described by his son, also an architect. Speer's mother hailed from a wealthy merchant family, and remained enamored of society life. She was a card-carrying Party member, having joined the Nazi Party in 1931 (a fact that she finally told her son about seven years later), and she also later gloried in dining at Hitler's side at the Berghof in the company of her famous son. However, his father—who was speechless upon meeting Hitler in 1938—never did join the Nazis that he so despised. He died in 1947, and Speers' mother died in 1952, while Speer was an incarcerated war criminal.

Albert had two brothers, Hermann and Ernst, who bullied him throughout his childhood. Although he supported his older brother Hermann financially after World War II, he refused his parents' entreaties to have Ernst evacuated from Red-Army-besieged Stalingrad during the war. Ernst died there.

The family moved permanently to their former summer home at Heidelberg in the summer of 1918, when Speer was 13—and defeat was closing in on Imperial Germany. He later recalled fondly that as a boy, he toured a giant Zeppelin airship in 1916.

Author Dan van der Vat noted that Speer had "The even temper of a man with no temper to lose," basically what Americans call "a cold fish." Comparing the two men, van der Vat noted: "Dr. Walter Schieber, head of armaments delivery with the rank of State Secretary, said after meeting Speer: 'He lacks Todt's human warmth. Even his heart is cool.'"[1] It was a telling characterization. His own fellow workers saw Dr. Speer as "Courteous, calm, friendly, humorous, and modest," stated Sereny. "He really talked the same way to everybody… One of his great talents was organizing… Another was delegating authority."[2]

Like Rudolf Hess—his fellow inmate at Spandau Prison for two decades—young Speer neither smoked nor drank, never danced, and married the first girl he dated.

Professor Speer himself seemed serene—perhaps resigned is a better word—concerning his place in the history of the Second World War. As he told *The Washington Post* in 1976, he should be remembered as, "One of the closest collaborators of Hitler. What I said at Nuremberg—that I was responsible for what happened to me—will stick with me,

Dr. Speer (center) smiles as Nazi Prime Minister of Prussia Hermann Göring (left) shares a beer with well-wishers in 1934. He was what Dr. Goebbels called "A Septemberling," i.e. a latecomer to the Nazi Party, and as such was initially looked down upon by all the other, longer established members of the Führer's tight-knit inner circle. He was known as a man of iron self-control: "The most determined of men at whatever he undertook," observed one onlooker. (HHA)

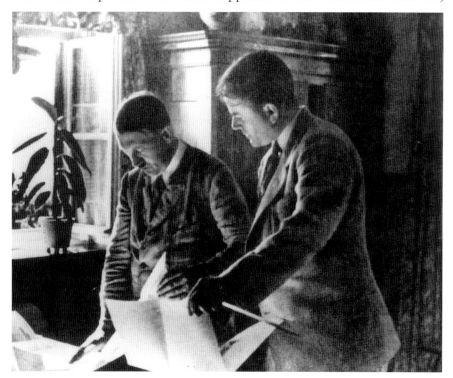

Speer (right) began his real career as the Führer's secondary architect, after the late Professor Paul Ludwig Troost died. Here, in 1934, he shows his patron, Hitler (left) some architectural plans at Haus Wahnfried on the Obersalzberg. (HHA)

A previously unpublished photo from the Eva Braun Hitler Albums showing the Führer and his entourage out for a stroll on the Obersalzberg sometime during 1933/34. From left to right are an unknown man, Hitler (carrying the hippopotamus hide dog whip that he affected until late 1934), his then liaison officer from Deputy Führer Rudolf Hess, Reichsleiter Martin Bormann, and Dr. Speer, Bormann's arch-rival within Hitler's intimate circle. (US National Archives, College Park, MD)

rightly. It will be my stamp. I hope that it will also be remembered that I was capable of three other things: to be an architect, a manager, and a writer."[3] Indeed, as his wartime work and postwar published memoirs attest, he *was* all three.

Fest mentions Speer's four lives: "Each self-contained and yet inconceivable without the preceding life: the beginning, when everything was possible, then the decade as an architect and Hitler's favorite minister, followed by more than 20 years of prison. And finally what he once called 'his posthumous life'—his success as an author and as a witness to an era."[4]

Ironically, Speer only entered the field of architecture because his politically liberal father decreed that he must. Boasting that he was "The best mathematician in my school," the future Nazi master builder put aside his own ambitions in order to study the discipline that both his father and grandfather had chosen. The 1923 German financial inflation dictated that he begin his architectural studies at the local Karlsrühe Institute of Technology. The financial stabilization of the following year allowed young student Speer to transfer to the more prestigious Technical University of Munich. In 1925, he transferred yet again, to the much esteemed Berlin Institute of Technology, where he became first the student, and then the personal assistant, of the famous architect-teacher Dr. Heinrich Tessenow.

In that capacity, Speer taught seminar classes three days weekly. It was his students, he claimed, who convinced him to hear Hitler speak at the university on December 5, 1930, and he was impressed. He recalled that, "I was carried along on the wave of the enthusiasm... It swept away any skepticism, any reservations."[5] After attending another Nazi rally, featuring Dr. Josef Goebbels as the main speaker, Speer joined the Nazi Party.

ATLANTIKWALL (ATLANTIC WALL)

"The West Wall shields the Reich,
while the Second Wall protects Europe"
Nazi propaganda slogan, 1942–44

After Germany had conquered Western Europe, there were some 5,000km (3,107 miles) of Atlantic coast to protect, especially after the United States entered the war, and an Allied landing became more likely. The result was fortification along the entire length of the coastline.

The Atlantic Wall was the largest construction project of the war, using 1.25 million tons of steel.

On December 14, 1941, Hitler ordered work on the Atlantic Wall to begin. Work was stepped up following commando raids by the Allies along the coast the following year.

Speer described the design of the fortification: "The larger ports were ringed with pillboxes, while the intervening coastal areas were only protected by observation bunkers at long intervals. Some 15,000 smaller bunkers were intended to shelter the soldiers during the shelling prior to an attack." Hitler planned that soldiers would then come out and fight in the open. He had input into the planning of the defenses, designing several bunkers and pillboxes himself. The overall design was completed by Todt, though much of the work was overseen by Speer after Todt's death. To simplify the work, a number of standardized bunker designs were created; once the type had been chosen, work could begin without delay as all the drawings and plans were already available.

The Organization Todt—charged with building the wall—put 500,000 men to work, according to the Schmieelkes, and by April 1943, 769,000 cubic yards of concrete had been emplaced, a signal accomplishment of the late Dr. Todt and his deputy Xaver Dorsch. Indeed, as noted by the Schmieelkes, the OT had jurisdiction over *both* the Army and the Navy in the overall construction of the wall from the very start, as had also been the case before with the Siegfried Line, during 1938–39. Much of the work was undertaken by slave laborers working for the OT.

In the autumn, von Rundstedt reported that the defense of France would stand or fall at the Atlantic Wall, but the wall itself was not yet adequate. He rightfully pointed out that once the line was broken, any remaining fortifications were completely useless, as they were all facing the sea. A month later Field Marshal Erwin Rommel visited to inspect the defenses, and he then implemented a program of beach obstacles. By the time the Allies invaded, 12,247 fortifications were complete, with half a million beach obstacles laid, but there were large stretches of the coast that were less densely protected, including Normandy. Speer concluded: "For this task we consumed—in barely two years of intensive building—17,300,000 cubic yards of concrete worth 3.7 billion DM. In addition,

A massive building site on the Atlantic Wall. Ironically, despite all of the construction, the combined Allied naval bombardment of June 6, 1944, breached the mighty Atlantic Wall in a single day. (*Signal* magazine)

The Commander in Chief West, von Rundstedt, on an inspection tour of the Atlantic Wall with staff officers. (CER)

Field Marshal Rommel (right) is greeted by a German naval officer (center) during one of his many inspection tours of the German-occupied Atlantic coastline of Europe. The posts at left were designed to destroy Allied gliders upon impact, and many, indeed, worked very well in that role in Normandy. (RC)

the armaments factories were deprived of 1.2 million metric tons of iron. All this expenditure and effort was sheer waste." This was because the Allies brought their own *Mulberry* ports with them, avoiding costly attacks on the well-defended ports, and bypassing the defenses.

The attack on D-Day breached the Atlantic Wall in a matter of hours. While the stronger defenses at the ports were a nuisance, the rest was redundant. The Atlantic Wall had tied up large reserves of manpower in static positions until the invasion, and the reliance on static defenses did not aid the Germans when the Allies attacked. The huge quantities of resources used in building the wall, particularly the steel, could have been used for guns, tanks, and ammunition. But even though that could have made a great difference to wartime production, the ultimate outcome of the war would most likely have been the same. Steven Zaloga places the blame squarely on Hitler: "The gnat bites by British commandos along the French and Norwegian coast provoked Hitler into a massive construction completely out of proportion to its tactical value. Hitler had a visceral enthusiasm for monumental fortification."[6]

A previously unpublished German view of the Atlantic Wall near Le Havre as faced by the Allies on D-Day, June 6, 1944. (LC)

Right: What the Atlantic Wall encompassed: from the North Cape on the Arctic Circle at upper right to the Spanish frontier at lower left, Hitler's Fortress Europe contained many individual fortress sites. At Cherbourg, there were 47,000 men, and 12,000 at St. Mâlo/Dinard. There were also 38,000 at Brest; 15,000 at Lorient, 25,000 at Quiberon Bay and Belle Isle; 35,000 at St. Nazaire; 14,000 at Le Havre; 10,000 at Boulogne; 9,000 at Calais/Cap Gris Nez; 12,000 at Dunkirk; 14,000 at Zeebrugge, and 8,000 in the Scheldt. (Atlantikwall Museum)

A previously unpublished photo of Speer (right) and Hitler (with binoculars) taking a break from a mountain walk on the Obersalzberg, accompanied by two SS security men, in 1935. In *Memoirs*, Speer said of the Führer's daily walk from the Berghof to the Mooslahnerkopf Teahouse: "We often walked in silence side by side, each dwelling on his own thoughts." Noted one observer of their symbiotic role with one another: "For the official political staffer … Speer was the Führer's friend and artistic colleague—a position that made him sacrosanct." Both Speer and Rudolf Hess agreed at Spandau that the Nazi Führer never had a real personal friend in the normal sense while they knew him. (EBH Albums, US National Archives, College Park, MD)

Young Speer almost moved to Afghanistan to work:

I very nearly became an official court architect as early as 1928. Aman Ullah, ruler of the Afghans, wanted to reform his country, and was hiring young German technicians with that end in view.

Josef Brix—Professor of Urban Architecture and Road Building—organized the group. It was proposed that I would serve as city planner and architect, and in addition as teacher of architecture at a technical school which was to be founded in Kabul … but no sooner was everything virtually settled—the King had just been received with great honors by President von Hindenburg—than the Afghans overthrew their ruler in a coup d'état.[7]

In 1928, Speer married Margarete Weber of Heidelberg (1905–87), who he had met in 1922, when she was 15 and he 16. His parents opposed their marriage as not suitable for a young man of his social standing in the community, but the pair married on August 8, 1928 at Berlin's famous Kaiser Wilhelm I Memorial Church, damaged by Allied bombing, but still standing today. Speer's parents did not attend, and did not invite the couple to stay with them until seven years later. Later, when at Spandau, Professor Speer recalled launching folding flatboats near the prison during their honeymoon. The couple had six children between 1934 and 1942: Albert Jr. (the architect of the 2008 Summer Olympics in Beijing), Hilde, Fritz, Margarete, Arnold, and Ernst.

In his memoirs, Dr. Ferry Porsche described Speer when he knew him in 1942: 'Tall, rapier-slim Albert Speer. Speer kept himself in good shape so that he conveyed the appearance of a sportsman. His dark hair was thinning in front, and his high forehead lent him an intellectual look."[8] Speer *was* a sportsman, as was his wife. Both liked to hike and ski in the mountains, pastimes that—on at least one occasion—earned Hitler's scorn: "I've always said that it's madness! With those long boards on your feet! Throw the sticks into the fire!"[9]

Dr. Porsche observed: "His gaze was direct and penetrating. I must say I was keenly aware of this man's brilliance and was impressed by it... Like everyone else at the time, Speer always wore a uniform..."[10]

Writer Omer Bartov said of Speer's character,

...his only true loyalty was to a psychotic murderer... Hitler was a second-rate painter but a first-rate politician, while Speer was a mediocre architect, but an extraordinary technocrat and organizer...

He may have been, as Fest argues, "A man of many abilities, but of no qualities," that is, a man who had to be pushed and motivated—perhaps even mesmerized, by some external force.

Two German Army soldiers on November 7, 1935 raise the new Reich War Flag, which Professor Speer was said to have personally designed. (Previously unpublished photo, HHA)

He may have been capable of serving a variety of masters and a wide range of political and ideological convictions and goals. Perhaps, indeed, he was a "man of the future—pragmatic, ambitious, without convictions," although there is enough evidence to show that he was a patriot, a nationalist, an anti-Communist, and an anti-Semite...[11]

Writer Phyllis G. Proctor noted: "Prior to the Nuremberg trial, all 22 defendants were given psychological tests. Although Speer's sanity was never questioned, the popular belief was that individuals so deeply involved in such a regime must certainly be mad. American psychiatrists were not particularly sympathetic to Speer—they viewed him as a morally weak-willed man with vaulting ambition who would allow nothing, not even the suffering and deaths of millions, to stand in the way of his goals."[12] I concur.

As noted by Joachim C. Fest in *Speer: The Final Verdict*:

Although he was one of the producers, Speer himself was undoubtedly gripped by these overwhelming emotions: seducer and seduced at the same time. "I was swept away," he admitted, adding that he would not have hesitated to follow Hitler "blindly ... anywhere."

He always insisted that the relationship that had developed between them had resembled that of "an architect toward an admired patron, rather than of a follower toward a political leader," but the one could not be separated from the other, particularly since "blind" devotion in matters of architecture would be nonsense. Not until much later did he realize that whenever the regime was accused of persecution or breaking treaties, he subconsciously began to search for justifications, and soon had joined the chorus of yes-men.[13]

In 1944, the British newspaper *The Observer* published an article on Speer which shows that even contemporary observers saw him as rather different to others of the Nazi Party. Speer actually discussed the article in his *Memoirs*:

What was really bothering me was ... that Bormann might show Hitler an article from ... *The Observer* in which I was described as a foreign body in the Party-doctrinaire works. I could easily imagine him doing so, and even the caustic remarks he would make. In order to anticipate Bormann, I myself handed Hitler the translation of this article, commenting jokingly on it as I did so. With considerable fuss, Hitler put on his glasses and began to read:

"Speer is—in a sense—more important for Germany today than Hitler, Himmler, Göring, Goebbels, or the generals. They all have, in a way, become the mere auxiliaries of the man who actually directs the giant power machine—charged with drawing from it the maximum effort under maximum strain... In him is the very epitome of the 'managerial revolution.'

The speaker's rostrum (top, center) and government benches (left and right) of the Nazi Reichstag, housed in the former Kroll Opera House in Berlin before and during the war. Dr. Speer was an elected member of the Reichstag from a Berlin suburb. Göring opens the session from the speaker's chair as Reichstag President in this previously unpublished view from his personal albums. (Hermann Göring Albums, Washington, DC)

"Speer is not one of the flamboyant and picturesque Nazis. Whether he has any other than conventional political opinions at all is unknown. He might've joined any other political party that gave him a job and a career. He is very much the successful average man, well dressed, civil, non-corrupt, very middle class in his style of life, with a wife and six children. Much less than any of the other German leaders does he stand for anything particularly German or particularly Nazi. He rather symbolizes a type which is becoming increasingly important in all belligerent countries: the pure technician, the classless bright young man without background, with no other original aim than to make his way in the world and no other means than his technical and managerial ability. It is the lack of psychological and spiritual ballast—and the ease with which he handles the terrifying technical and organizational machinery of our age—which makes this slight type go extremely far nowadays... This is their age; the Hitlers and Himmlers we may get rid of, but the Speers— whatever happens to this particular special man—will long be with us."

Hitler read the long commentary straight through, folded the sheet, and handed it back to me without a word, but with great respect.[14]

A previously unpublished photo of former Secretary of State in the Propaganda Ministry and Gauleiter of Lower Silesia Karl Hanke (right) at Hitler's principal wartime military Fuhrer Headquarters, the Wolf's Lair at Rastenberg, East Prussia, accompanied by other Gauleiters. They were the effective domestic rulers of the Third Reich under their boss—and Speer foe—Reichsleiter Martin Bormann throughout the war. Speer had many interactions with them, not always pleasant for either him or them. Hanke had a falling out with his boss Dr. Goebbels over being in love with the propaganda minister's wife, Magda, and was dismissed from his personal staff as a result. As Gauleiter of Fortress Breslau in 1945 against the Red Army, Karl Hanke impressed Hitler to the extent that he was named Reichsführer-SS, succeeding the traitor Himmler. Hanke escaped from Breslau in a helicopter, but was captured by Czech Red Partisans and killed while attempting to flee, disguised as an ordinary SS man. (HHA)

Within his craft, Fest added: "During these years, Speer had also been working as a freelance architect. His office had expanded steadily, and he was almost stupefied by the never-ending flood of inquiries, commissions, journeys, and administrative duties, often coming home late in the evening, 'speechless with exhaustion.' To begin with, he had refused to accept a fee for his official work, but he increasingly got into difficulties. Only toward the end of 1935, when Göring assured him with his constantly cheerful greed, 'They're all nonsense, your ideals. You've got to make money!' did Speer accept a fee of 3,000 Reich Marks for his work up until then."[15]

In 1932, Speer's first patron-connection within the Nazi Party—Karl Hanke (later a *Gauleiter*)—brought him to Dr. Goebbels' attention to renovate the Berlin district headquarters, and, still later, to redo the Propaganda Ministry Building in the capital after the Nazis took office under Hitler as Reich Chancellor. Pleased with the work, Dr. Goebbels recommended Speer to Hitler to help Führer architect Dr. Paul Ludwig Troost renovate the Old German Reich Chancellery in Berlin. Since Dr. Troost was a Munich architect—and not as familiar with the Berlin market as was the younger man—Professor Speer's help proved valuable.

Speer's best-known contribution to the Old Reich Chancellery (ORC) was the famous balcony from which Hitler later reviewed parades and received masses of well-wishers following his diplomatic and military triumphs. In the course of this work, Hitler and Speer became close friends and associates, especially as Hitler had been a failed student artist in his own youth, and fancied himself as a world-class architectural thinker and builder.

Dr. Speer's best-known works are the mammoth Nazi Party buildings at the national parade site of Nuremberg, in southern Germany. Another of his most famous works was the massive New German Reich Chancellery in Berlin, which debuted in January 1939, with its 1943 underground air-raid shelter that later became the Führer's last wartime command headquarters.

On January 30, 1937—the fourth anniversary of his being appointed Reich Chancellor—Hitler named his young protégé as General Building Inspector for Berlin, with the specific task of rebuilding the traditional city as the new, future, world capital of Germania, with the rank of State Secretary in the Reich Cabinet, which meant that he was, in effect, serving as the Führer's own deputy in all matters architectural, reporting to him alone.

The young man of 32 had arrived. He and his patron completely meshed, asserted Fest, because Hitler was "Always ready to take the

A jovial Professor Albert Speer in 1941 on the Obersalzberg, as photographed personally by Eva Braun (1912–45). They became fast friends, and enjoyed skiing and hiking outdoors in the beautiful Bavarian mountains and countryside. (Previously unpublished photo, US National Archives)

THE SAGA OF THE ME 262

On May 22, 1943 German Luftwaffe General of the Fighters Adolf Galland was invited by designer-builder Professor Willi Messerschmitt to test-fly the prototype of the new Me 262, the world's first operational jet-powered fighter aircraft. After the war, General Galland described the difficulties he faced with the new aircraft, but he was very excited about its flight characteristics, handling, and speed. General Galland referred to the ensuing events as "the jet fighter tragedy": the Third Reich flew the prototype a *full two years before the end of the war*, and yet couldn't get it into mass production in time to stave off defeat!

Following his test flight, Galland submitted a report, with suggestions that preparations be made for mass production, and the existing prototypes be tested in the meantime, to Göring, and copied to Erhard Milch, one of Hitler's favorite Luftwaffe field marshals. Göring and Milch both accepted the suggestions. Galland was delighted, and felt sure Hitler would sanction the decision.

But Hitler refused to give his approval. Citing previous disappointments with promises of innovations and improvements by the Luftwaffe, he would not be rushed into approval of the Me 262. He allowed tests to continue, but forbade any preparation for mass production. This delayed production of the Me 262 for a further six months after a previous delay of two years, the latter as a result of an order in the fall of 1940 which had stopped all research development. Meanwhile, the Allies were achieving growing superiority in both quantity and quality of fighters.

Speer was advised of the situation by Milch in September when the order from Hitler to halt preparations for production was received. They chose to continue, but not with the priority that they would have wished. Three months later, Milch and Speer were summoned to see Hitler, who had changed his mind after seeing something in the British press about experiments with jet planes. He wanted as many Me 262s built as quickly as possible. The best that could be done was 60 a month from July 1944, increasing to 200 a month from 1945. However, Hitler wanted to use the plane as a bomber. The specialists

From left to right (foreground), Luftwaffe Field Marshal Erhard Milch, Minister Speer, and (far right) Dr. Ferdinand Porsche. Of Milch, Speer wrote in 1970: "Especially during the early months [as armaments minister], his advice became indispensable; out of our official relationship there grew a cordial friendship which has lasted to the present." They testified on each other's behalf at Nuremberg. (CER)

The Messerschmitt Me 262, the world's first operational jet, produced by German industry in 1942, didn't enter aerial combat until July 1944. (CER)

Above: German General of the Fighters Adolf Galland in 1944. Noted Professor Speer: "Two of the most successful Air Force officers—[Werner] Baumbach and Galland—worked with me during the last days of the war developing a weird plan for laying hands on the most important members of Hitler's entourage and preventing them from committing suicide." Despite Speer's postwar assertions, their joint plans came to nothing. (CER)

hoped to persuade him, but he was adamant, insisting that all weapons be removed to allow more bombs to be carried; the speed of the planes meaning that they would not have to defend themselves. Speer and the Luftwaffe were dismayed. As bombers, the Me 262 would be insignificant, but as a fighter it could have had a tremendous effect against the American bombers now targeting German cities.

In May 1944, Hitler met with his Luftwaffe chiefs, Speer, Milch, Karl-Otto Saur, Colonel Petersen and several others involved with the Me 262, but not General Galland. Discussion moved onto Saur's production program for the Me 262 fighter. Milch had gambled, believing that only this plane could save the air war. Hitler was furious to hear that the planes were being produced as fighters, not fighter-bombers, after his decision of the previous year, and that his orders had been disobeyed by his most trusted officials. On the spot, the Führer modified the plans for the plane. His officials were shocked, but Milch tried once more to persuade him. A row ensued, which essentially ended Milch's career. On May 29, Göring ordered the plane transferred from Galland's office to that of the head of the Bomber arm.

The Me 262 affair clearly shows up the fact that by this time, Göring no longer stood up to Hitler. Hitler was criticising the Luftwaffe severely, and Göring's reaction was to simply submit to his wishes, which in the end doomed the Luftwaffe. Athough Göring agreed with Galland and Milch as to the potential of the plane, he was reluctant to commit to the Me 262 as a fighter. He was eventually persuaded to protest to Hitler, but the orders of late May 1944 show that he was defeated. He then would have had to mediate with the unhappy parties who believed in the Me 262 as a fighter. As a result of the affair, Göring's prestige within the Luftwaffe was damaged.[16]

By June 1944, the Luftwaffe was desperate to use the new aircraft against the bombers, but Hitler was still refusing to let them be used as fighters. The General Staff also had been counting on the new fighter to make a difference in the air war, and so "everyone who could claim any knowledge of the subject at all put in a word and tried to change Hitler's mind: Jodl, Guderian, Walter Model, Sepp Dietrich, and, of course, the leading generals of the Air Force."[17] Eventually Hitler forbade any further discussion on the topic.

Galland, with the support of Speer, had continued very small-scale tests of the Me 262 as a fighter. At last, in October 1944, Göring ordered him to form a jet fighter unit from the test commandos. The suggestion had come from Himmler, and his support was needed for Göring to be able to put it into action. The successes of these fighter units finally persuaded Hitler that the plane was an excellent fighter.

The following March, however, Hitler ordered that the fighter-bomber Me 262 be rearmed as a fighter as quickly as possible. But, by then, it was of course too late to stop the onward rush of the Allied armies on the ground, much less their fleets of bombers and fighters overhead.

most eccentric ideas seriously and put them into effect with that fearlessness with which he etched himself so indelibly on the world's memory."[18] Whatever else can or cannot be said for Adolf Hitler, he never did anything by halves, and thus Speer emerged as the handmaiden of his will in many of his better-known projects, such as their joint buildings, some of which can be seen in films of the time, and other which can only be viewed as models, as they were never actually built. Indeed, Fest named over 40 "Führer cities" that were slated for rebuilding by this dynamic duo, aside from Berlin.[19]

By 1940, Dr. Speer had emerged as a major power player within Nazi Germany with but a sole formidable enemy—Martin Bormann—who would fight him tooth and nail until the very end of the war in 1945.

"None of the bigwigs came to our house on the Berg [Obersalzberg, the mountain retreat]," recalled Speer later. But the biggest one of all did: during his daily walks while on the Obersalzberg during the war, Hitler would drop in at the Speers' nearby country home and play with their children. Daughter Hilde is being held, while one of the sons stands aloof at left. Reportedly, they never warmed up to Hitler. (EBH albums, US National Archives)

A trio of Berghof families pose for a group picture with their Führer during the war: the Speers, Brandts, and Bormanns, with all their combined children. Dr. Brandt is at the far left, top row; Professor Speer is to the right of him, and Frau Anni Brandt is at Dr. Speer's left. Next to her is Dr. Theo Morell, Hitler's personal physician. The Führer is in the center, holding hands with two of the children, and Martin Bormann is at the far right. Standing above Bormann is Hitler's longtime personal photographer, Reich Photo Reporter Professor Heinrich Hoffmann. The other women are servants and Führer secretaries, except Eva Braun, standing to the right and just above Hitler, and Frau Morell just behind her at left. (HHA)

Fest asserts that Speer, "sided with the radicals in favor of war" in 1939 (the "war party" consisted of Himmler and Foreign Minister Joachim von Ribbentrop, while the "peace" faction included Göring and Dr. Goebbels, who believed that they had everything to lose by war and nothing to gain). Hitler opted for war, so war it was.[20]

With the coming of the war, ironically, Speer's rising star in Nazi Germany started to crest, since his role as the major domo in the building sphere gained him no laurels at Hitler's military conference tables at the various Führer Headquarters spread out across German-occupied Europe.

It was as Dr. Fritz Todt's successor that Professor Speer was set to take his place on the world stage, becoming as familiar to western newsreel audiences as he was at home in the Reich. In effect, by reversing Göring's alleged earlier blunders made during the latter's Four Year Plan economic dictates of 1936–42, Speer returned armaments and war production to the German barons and captains of industry by ousting incompetent Nazi Party bureaucrats. These "new" men were those of German big business, who knew what they were doing. This was the essence of his brilliance.

CHAPTER SEVEN

NAZI ROCKETRY, 1933–45

> "This is the decisive weapon of the war! What encouragement to the home front when we attack the English with it! Speer, you must push the A-4 as hard as you can! Whatever labor and materials they need must be supplied instantly."
>
> Adolf Hitler, July 7, 1943

> "Gigantic effort and expense went into developing and manufacturing long-range rockets which proved to be—when they were at last ready for use in the autumn of 1944—an almost total failure. Our most expensive project was also our most foolish one. Those rockets—which were our pride, and for a time my favorite armaments project—proved to be nothing but a mistaken investment. On top of that, they were one of the reasons we lost the defensive war in the air."
>
> Dr. Albert Speer, 1970

General Karl Emil Becker (1879–1940), who, according to Neufeld, "Initiated the German Army rocketry program in the early 1930s. He was convinced that liquid-fuel rocketry provided the key to a devastating new secret weapon: the long-range, ballistic missile … and had begun to investigate the revival of the rocket as a weapon." He was proven right. "The Army had made a massive investment in ballistic missile technology since the mid-1930s." Formerly the Dean of the Faculty of Military Technology at Berlin's Technical University, General Becker in 1939 was Chief of the Army's Weapons Procurement Office and—according to author Adam Tooze—"Germany's leading ballistics expert." During a trip on Hitler's Special Train Amerika on April 8, 1940, ordnance specialist Erich "Cannon" Müller convinced Hitler that the Third Reich needed Dr. Fritz Todt as overall armaments production tsar instead, and—as a direct result—General Becker shot himself, having lost his Führer's confidence. (CER)

The initial usage of rocketry in warfare occurred in China, and then in the European wars of 1250–1400.[1] The rocket was then mainly eclipsed by the cannon, but development continued in India, and rockets were used there against the British in the late 18th century. Britain then adopted and improved rockets, using them from 1805 onwards. By the mid-19th century, the Americans, Austrians, and Russians were all using rockets, but they were surpassed by the 19th-century revolution in artillery. World War I was an artillery war, and the rocket saw limited use.[2]

In 1919, the American physicist Dr. Robert H. Goddard published his important work, *A Method of Reaching Extreme Altitudes*, followed by Professor Hermann Oberth in 1923 with *The Rocket and Space Travel*. In 1927, the Society for Space Travel was established in the Weimar Republic. Forced by the strictures of the 1919 Treaty of Versailles to look for new weaponry not condemned by it, the German Army Weapons Department embraced the "novel" idea of rocketry for the future.

Writing in 1952, Dr. Walter Dornberger recalled: "In the spring of 1930, after finishing my technical studies, I was appointed to the Ballistics Branch of the Army Weapons Department... As we didn't succeed in interesting heavy industry, there was nothing left to do but to set up our own experimental station for liquid-propellant rockets at the department's proving ground in Kümmersdorf near Berlin... The Experimental Station West was ... between the two Kümmersdorf firing ranges, about 17 miles south of Berlin, in a clearing in the open pine forest of the Province of Brandenburg."[3]

On December 21, 1932, the first Kümmersdorf rocket test firing occurred. Present was 19-year-old Dr. Wernher von Braun, who had joined Dr. Dornberger's staff the previous October 1. He was a distant cousin of Hitler's mistress, and later wife, Eva Braun Hitler. Of him, Dr. Dornberger wrote in 1954: "I'd been struck ... by the energy and shrewdness with which this tall, fair, young student with the broad massive chin went to work, and by his astonishing theoretical knowledge... He grasped the problems ... a refreshing change from most of the men at the place."[4]

The A-1 (Aggregat 1) rocket was the first complete missile fired, followed by the initial two A-2 rockets launched from the Isle of Borkum in December 1934, after Hitler had taken office as German Reich Chancellor.

In April 1936, Göring's Luftwaffe officer Lt. Col. Wolfram von Richthofen (the late Baron Manfred von Richthofen's cousin) led a delegation to ask the then Army Gen. Albert Kesselring for the use of its Experimental Station at Peenemünde, located at the mouth of the Peene River on the Baltic coast, as a test site preserve for the future development of German rocketry. At the time, von Richthofen was Chief of the Air Ministry's Development Division, and Kesselring was Chief of Aircraft Construction.

Ironically, the Peenemünde site was suggested, reportedly, by von Braun's own mother, and the inlet was perfect for Dr. Dornberger's needs: "We had to fire out to sea, and to observe the entire trajectory from land." General Kesselring approved, construction began, and 120 engineers and scientists were duly assigned to the new site; transfer from Kümmersdorf West occurred in August 1939. Eventually over

A meeting on July 23, 1930 of some of Weimar Republican Germany's early rocket pioneers. At far left is Rudolf Nebel, a founder of the later *Raketenflugplatz* (Rocket Launch Site), "the most important amateur rocket group," according to one missile authority. At center, to the right of the rocket, is Hermann Oberth, touted as "the pioneer of the Weimar space flight movement," who demonstrated his rocket engine in Berlin. Klaus Riedel—chief designer of the group— is holding the smaller rocket on the long stick in the forefront. Behind him and to the right is the young Wernher von Braun. (CER)

DR. WALTER ROBERT DORNBERGER (1895–1980)

Born in the Ruhr Valley of western Germany in 1895, Dornberger joined the Imperial German Army in World War I. In October 1918, he was captured by the US Marines and was in a French prisoner-of-war camp in solitary confinement until 1919 due to repeat attempts to escape.

After returning home, he studied physics and engineering. In 1930, Dornberger received an MS degree in engineering from the Technical College of the University of Berlin. In 1932, Dornberger was put in charge of a group of a scientists working on rocket concepts. In 1935 he was given an honorary doctorate. In 1937, Dornberger moved his rocket command to Peenemünde where the first successful test of the V-2 was carried out.

At the end of the war he surrendered himself to American troops, offering his services to the US as a rocket scientist. He spent two years in a British prisoner-of-war camp, before going to the US in 1947, where he worked for the US Army and Air Force. Like more than 100 other German-born rocket scientists, he became an American citizen. Dornberger died in West Germany in 1980 at the age of 84.[5]

A famous photograph of General Dornberger (left) and Dr. von Braun. The general wears the Nazi War Merit Cross with Swords at his throat. At his 1944 meeting with Gestapo (Secret Police) chief SS General Heinrich Müller, the latter told Dr. Dornberger, "You are regarded as our greatest rocket expert." (CER)

In a photograph taken during the October 1940 visit of Dr. Fritz Todt to Peenemünde, young Dr. Wernher von Braun is seen at center in a dark civilian suit. Colonel General Friedrich Olbricht is at the left, cradling a cigar in his left hand. He was shot on July 20, 1944 as a traitor to Hitler. (CER)

Above: Von Braun (third from left), in his role as Army Experimental Station Peenemünde's technical director, explains what is going on at a wartime rocket launching to Wehrmacht and Kriegsmarine/Navy officers. (Smithsonian Air and Space Museum, Washington, DC)

18,000 people were employed in the area. These included 1,500 scientists and engineers, and 8,000 specialists. Planning for the A-3 rocket had already begun in July 1936.

Accompanied by Göring and Defense Minister Gen. Werner von Blomberg, Hitler first visited Kümmersdorf in October 1933. In March 1939, he visited Kümmersdorf West a second time, accompanied by Army Commander in Chief Col. Gen. Walther von Brauchitsch, Gen. Emil Becker, Deputy Führer Rudolf Hess, and the latter's assistant, Reichsleiter Martin Bormann. Dr. Dornberger said of Hitler's visit:

> I still don't know whether he understood what I was talking about. Certainly, he was the only visitor who had ever listened to me without asking questions.
>
> We went over to the old test stand to witness a test run of a rocket motor developing a thrust of 650 pounds. The horizontally suspended combustion chamber was ignited. When the harsh roar of the pale blue jet of gas—concentrated in a narrow stream with the supersonic shock waves clearly delineated in colors of varying brightness, caused a painful vibration in our eardrums in spite of thick wads of cotton wool—his expression did not change.

Dr. Dornberger (left) and his top engineers in Vienna during the summer of 1942. The other three are, from left to right, Col. Leo Zanssen, Dr. Walter Thiel, and Dr. Wernher von Braun. (CER)

A V-1 flying bomb as it looks today, preserved at the Smithsonian National Air and Space Museum on the Mall in Washington, DC. Instead of using it against the Allied invasion beaches in Normandy, Hitler decided that it would be a weapon better deployed against London as a political terror expedient designed to overthrow the Churchill War Cabinet, and thus get Great Britain out of the war. (Blaine Taylor)

Nor did the next demonstration—with a vertically suspended motor developing a thrust of 2,200 pounds—draw a single word from him. Hitler watched this test run, standing behind a protective wall, from a distance of only 30ft.

On the way to one of the assembly towers at the third test stand, I told him about our work at Peenemünde, and of the results we had achieved. The Führer of the German people walked on beside me, staring ahead and holding his tongue. In the assembly tower, we had assembled a cutaway model of the A-3 horizontally on low wooden

trestles. One could see—through slits and holes in the thin outer sheet-metal skin—the standpipes, valves, tanks, and the rocket motor, and observe the flow of propellants and the control processes. To make it easier to understand, related components were painted the same color. While Hitler was looking into the rocket, von Braun gave technical explanations, describing how the entire system worked. Hitler examined the machine very closely from all sides, and finally turned away, shaking his head...[6]

Hitler also viewed the A-4 and A-5 models. Dornberger continued:

During lunch in the mess, I sat diagonally opposite Hitler... He asked casually how long it might take to develop the A-4, and about its range. When I named the long peacetime standard periods, he answered with a brief nod. Finally, he wanted to know whether we could use steel sheeting instead of aluminum [for the surrounding skins of the rocket bodies]. When I did not reject this possibility, but emphasized that it would cause delay, he looked past me with an absent smile and uttered the one word of appreciation that was to be vouchsafed to us: "Well, it was grand!"...

I simply could not understand why this man ... could not take in the true significance of our rockets... He could not fit the rocket into his plans... The engineering spectacle had no doubt fascinated him to some extent... It was this, perhaps, that made him let us go on with the work.[7]

Dornberger was right, for Hitler's war was slated to begin but a few months later, and he was counting above all else at that point on a

Left to right: Gen. Walter Dornberger, German Army C-in-C Col. Gen. Walther von Brauchitsch, other officers, and Dr. Wernher von Braun in dark civilian suit at right. Following two years in a British POW camp, Dr. Dornberger went to the US and became an American citizen, dying at the age of 84 in 1980. (US National Archives)

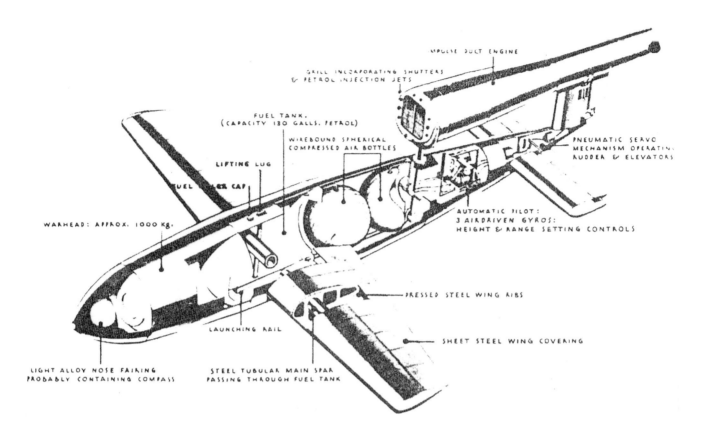

IMPULSE DUCT ENGINE

GRILL INCORPORATING SHUTTERS
& PETROL INJECTION JETS

FUEL TANK.
(CAPACITY 130 GALLS. PETROL)

WIREBOUND SPHERICAL
COMPRESSED AIR BOTTLES

LIFTING LUG

FUEL FILLER CAP

PNEUMATIC SERVO
MECHANISM OPERATING
RUDDER & ELEVATORS

WARHEAD: APPROX. 1000 Kg.

AUTOMATIC PILOT:
3 AIRDRIVEN GYROS:
HEIGHT & RANGE SETTING CONTROLS

PRESSED STEEL WING RIBS

LAUNCHING RAIL

SHEET STEEL WING COVERING

LIGHT ALLOY NOSE FAIRING
PROBABLY CONTAINING COMPASS

STEEL TUBULAR MAIN SPAR
PASSING THROUGH FUEL TANK

Cutaway diagram of a V-1, called a "doodlebug" and a "buzz bomb" by the British because of the pulsating noise of its engine, and whose country it hit during 1944–45. According to the US Air Force, "The V-1 was a pilotless aircraft developed by German Air Force engineers during World War II for use against Allied cities such as London and Antwerp. It was powered by an air-breathing pulse-jet engine using aviation gasoline as its fuel. The warhead (in the nose) contained nearly a ton of high explosives. The technical designation for the V-1 was the Fi-103 or FZG 76. The V-1 was launched from specially prepared catapult ramps. After it had flown a preset distance, a timer shut off the engine, causing the V-1 to dive. The missile's relatively low speed made it easy prey for radar directed antiaircraft artillery and fighter aircraft." (LC)

short, lightning victory, not the longer, protracted war that developed in the actual event. At this point, it was Dr. Dornberger's hope that the A-4—Dr. Goebbels's later V-2—could be mass-produced by December 1943. Speer now became involved. He recalled:

Even since the winter of 1939, I had been closely associated with the Peenemünde development center, although at first all I was doing was meeting its construction needs. I liked mingling with this circle of nonpolitical young scientists and inventors headed by Wernher von Braun, 27 years old, purposeful, a man realistically at home in the future.

It was extraordinary that so young and untried a team should be allowed to pursue a project costing hundreds of millions of Marks and whose realization seemed far away. Under the somewhat paternalistic direction of Col. Walter Dornberger, these young men were able to work unhampered by bureaucratic obstacles and pursue ideals which at times sounded thoroughly utopian...

Whenever I visited Peenemünde, I also felt—quite spontaneously—somehow akin to them. My sympathy stood them in good stead when in the late fall of 1939 Hitler crossed the rocket project off his list of urgent undertakings, and thus automatically cut off its labor and materials.

By tacit agreement with the Army Ordnance Office, I continued to build the Peenemünde installations without its [official] approval—a liberty that probably no one but myself could have taken.[8]

This proved not to be the final time Dr. Speer chose to circumvent Hitler's orders. As we shall see, he was to make another such crucial decision later on, in regard to the so-called German atomic bomb. Another man who protected the Peenemünde project, this time from German draft calls on its personnel, was the later Field Marshal von Brauchitsch, from September 1939 until Hitler fired him as Commander in Chief Army in December 1941 following the defeat by the Red Army at Moscow.

On October 3, 1942, Dr. Dornberger gave the excited command "Rocket away!" and there occurred the first successful test launching of the A-4 (V-2) rocket at Peenemünde, which was shortly to be taken over by "the blacks," as the SS were called. By then, Dr. von Braun had become Chief of Engineering.

Remembered Dr. Dornberger in 1954, "This was the final verdict... I had borne sole responsibility for more than 10 years." The new V-2 weighed 13 tons was a multi-stage rocket that was launched vertically upward, then tilted away at an angle of 50 degrees for maximum range eastwards, and traveled for 120 miles. "Supersonic speed was achieved for the first time by a liquid-propelled rocket at six miles high, at twice the speed of sound."9

The painting *Fighting Planes in Action* by Gary Sheahan depicts an American P-47 Thunderbolts (top) and a trio of RAF fighters below shooting down a V-1 "flying bomb" over Dover in June 1944 in this US Army Combat Art Collection piece. (US Army Combat Art Collection, Washington, DC)

Above: A V-1 caught in a snapshot in flight over England, 1944. (US National Archives, College Park, MD)

Above: A V-2 rocket (left) as it appeared in 1992 on display at the Smithsonian National Air and Space Museum. According to the US Army Ordnance Museum: "The V-2 rocket first appeared during the spring of 1944, just prior to the Allied invasion of Normandy. Some 4,300 V-2s were operationally launched during World War II. About 1,500 were directed against England, and over 2,100 were aimed at Antwerp, Belgium, which had become the principal port of entry for Allied supplies during the spring of 1944. About 6,590 others malfunctioned after launch, and did not reach their targets." (Blaine Taylor)

In a meeting with Minister Speer on January 8, 1943, Dornberger and von Braun advanced a plan to build a new rocket-launching site on the Channel coast. Dr. Speer told them that Hitler had disapproved of their March 1942 memo on the subject, but that he—the Minister of Armaments—would secretly build it anyway, in hopes that the Führer would come round to the idea later.

In the future, the two rocket men would report to Dr. Speer, and now also to Director Gerhard Degenkolb, Chairman of the Locomotives Special Committee within the Arms Ministry, who would serve as their liaison with German heavy industry. As Speer told Dornberger: "He is to set up an A-4 production committee. He has shown such drive and ruthlessness that he can manage the seemingly impossible without any high priorities, purely on the power of his name and personality."[10]

Regarding him, Dr. Dornberger noted: "Degenkolb inspected Peenemünde. He saw an A-4 about to take off and in flight. The effect

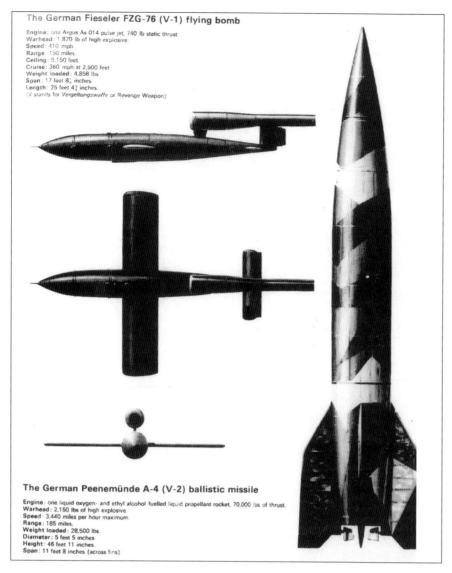

The German Fieseler FZG-76 (V-1) flying bomb

Engine: one Argus As 014 pulse jet, 740 lb static thrust.
Warhead: 1,870 lb of high explosive.
Speed: 410 mph.
Range: 150 miles.
Ceiling: 9,150 feet.
Cruise: 360 mph at 2,500 feet.
Weight loaded: 4,858 lbs.
Span: 17 feet 8¼ inches.
Length: 25 feet 4¾ inches.
(V stands for *Vergeltungswaffe* or Revenge Weapon)

The German Peenemünde A-4 (V-2) ballistic missile

Engine: one liquid oxygen- and ethyl alcohol-fuelled liquid propellant rocket, 70,000 lbs of thrust.
Warhead: 2,150 lbs of high explosive.
Speed: 3,440 miles per hour maximum.
Range: 185 miles.
Weight loaded: 28,500 lbs.
Diameter: 5 feet 5 inches.
Height: 46 feet 11 inches.
Span: 11 feet 8 inches (across fins)

A comparison of the German V-1 and V-2 rockets, in terms of both statistics and relative sizes. According to the US Army Ordnance Museum: "The V-2—the first successful inertially guided ballistic missile—used three gyroscopes and accelerometers as the main components of the guidance system. Control was accomplished by the use of jet valves in the exhaust stream, and air vanes for use in flight. It was a liquid-propulsion missile using ethyl alcohol for fuel and liquid oxygen (Lox) as the oxidizer. The rocket is launched vertically, and can be radio-controlled up to 65 seconds, after which time the rocket was in free flight. Historically, the V-2 was mankind's first reach into outer space, and heralded the space age." (LC)

was the same upon him as upon any other human being. The first impression was overpowering."[11]

Another more powerful opponent within the overall Speer Ministry had arisen however, Karl Otto Saur, who accosted Dr. Dornberger: "I suppose you think that you've struck it lucky today, now that you've got your special committee at last, but don't be too sure. Trees have never grown so high as heaven yet! You haven't yet convinced me or won me over, any more than you have the Führer!"[12]

Degenkolb attempted to take Peenemünde away from the Army and turn it into a stock company with limited liability and run by a board, but Dornberger successfully kept it under Col. Gen. Fritz Fromm, commander of the Home or Reserve Army; and a Long-Range Bombardment Development Commission was set up within Speer's Ministry of Munitions in February 1943, a year after he'd taken office as Minister. By December 1943, Degenkolb wanted to have 300 A-4s in completed production monthly: "He acted like a burly, endlessly threatening, and

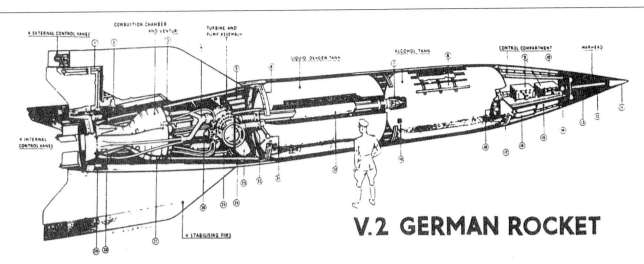

V.2 GERMAN ROCKET

4 EXTERNAL CONTROL VANES
COMBUSTION CHAMBER AND VENTURI
TURBINE AND PUMP ASSEMBLY
LIQUID OXYGEN TANK
ALCOHOL TANK
CONTROL COMPARTMENT
WARHEAD
4 INTERNAL CONTROL VANES
4 STABILISING FINS

1 CHAIN DRIVE TO EXTERNAL CONTROL VALVES.

2 ELECTRIC MOTOR.

3 BURNER CUPS.

4 ALCOHOL SUPPLY FROM PUMP.

5 AIR BOTTLES.

6 REAR JOINT RING AND STRONG POINT FOR TRANSPORT.

7 SERVO-OPERATE ALCOHOL OUTLET VALVE.

8 ROCKET SHELL CONSTRUCTION.

9 RADIO EQUIPMENT.

10 PIPE LEADING FROM ALCOHOL TANK TO WARHEAD.

11 NOSE PROBABLY FITTED WITH NOSE SWITCH OR OTHER DEVICE FOR OPERATING WARHEAD FUZE.

12 CONDUIT CARRYING WIRES TO NOSE OR WARHEAD.

13 CENTRAL EXPLORER TUBE.

14 ELECTRIC FUZE FOR WARHEAD.

15 PLYWOOD FRAME.

16 NITROGEN BOTTLES.

17 FRONT JOINT RING AND STRONG POINT FOR TRANSPORT.

18 PITCH AND AZIMUTH GYROS.

19 ALCOHOL FILLING POINT

20 DOUBLE WALLED ALCOHOL DELIVERY PIPE TO PUMP.

21 OXYGEN FILLING POINT.

22 CONCERTINA CONNECTIONS.

23 HYDROGEN PEROXIDE TANK.

24 TUBULAR FRAME HOLDING TURBINE AND PUMP ASSEMBLY.

25 PERMANGANATE TANK (GAS GENERATOR UNIT BEHIND THIS TANK).

26 OXYGEN DISTRIBUTOR FROM PUMP

27 ALCOHOL PIPES FOR SUBSIDIARY COOLING.

28 ALCOHOL INLET TO DOUBLE WALL.

29 ELECTRO HYDRAULIC SERVO MOTORS.

A cutaway diagram of a V-2 rocket. James McGovern in his 1966 work *Crossbow and Overcast* stated: "On the night of June 12, 1944 at 9p.m., General Jodl … sent a message to Col. Max Wachtel, commanding Flak Regiment 155 (W) in northern France. Operation *Rumpelkammer* (Lumber Room), the codename for the V-weapons' firing, despite its technical shortcomings, had to begin now." Added writer I. M. Baxter in his article "Hitler's Vengeance Weapons": "As Wachtel hastily began preparing his flak regiments for attack, all along northern France, thousands of laborers were still building and erecting mobile sites, and massive concrete buildings, both for the V-1s and V-2s." (Smithsonian National Air and Space Museum, Washington, DC)

dreaded slave driver."[13] In the midst of this came word from Hitler's FHQ: "The Führer has dreamed that no A-4 will ever reach England!"

Now a totally unexpected development occurred: although designed after the A-4/V-2, the V-1 flying bomb became operational before it. Professor Speer promised to remove Degenkolb if necessary, and even the glum Saur came aboard—but not Hitler, yet.

It was Professor Speer who informed Dr. Dornberger via telephone from FHQ that he'd been promoted to the rank of major general—and also, later, that a meeting had been scheduled with Hitler for July 7, 1943 for Dornberger to make a renewed sales pitch for the A-4 rocket once and for all, along with von Braun. They flew to Rastenburg in an He 111 bomber, bringing with them a film of the October 3, 1942 successful A-4 launch. Hitler himself had never visited Peenemünde, although Speer, Göring, Dr. Goebbels, and many others had. Dr. Dornberger recalled preparing for the meeting:

We had packed everything—the film, the model of the big firing bunker on the Channel coast, the little wooden models of vehicles, the colored sectional drawings, the organizational plans, the manual for field units, the trajectory curves.

I had not seen Hitler since March 1939… He had never seen—even in a movie—the ascent of a long-range rocket, never experienced the

SS GENERAL DOCTOR HANS KAMMLER (1901–45?)

Doctor of Engineering SS Gen. Hans Kammler was born on August 26, 1901, and "supposedly fell as a divisional commander during the battle for Berlin" in 1945.[14] He was a commander in the SS from 1938. Until the end of 1941 he was government construction director in the Luftwaffe. He then became SS *Oberführer* (Senior Colonel) and head of Office Group C in the new Main Economic Administration Office, making him responsible for concentration camp buildings, among other things. In 1942, he became Brigadeführer (Brigadier). In 1943, he oversaw the demolition of the Warsaw Ghetto after the Warsaw Uprising.

Kammler was in charge of constructing facilities for secret weapons, such as the Me 262 and the V-2, and in mid-1943, he was charged with moving the Peenemünde facility underground following Allied bombing raids. The factory at Nordhausen was constructed under his supervision, mainly by labor from concentration camps.

Speer made Kammler his representative for "special construction tasks," hoping that Kammler would work with his Armaments Ministry's main construction committee, but in March 1944, Göring appointed him as his delegate for "special buildings" under the fighter aircraft program, giving him greater power, and removing much of Speer's influence.

In August 1944, Kammler was promoted to SS major general, and two days later Himmler put him in charge of the V-2 program, responsible directly to him, and therefore not to Speer's Armaments Ministry. General Kammler was also subordinate to SS Gen. Hans Jüttner, Himmler's chief of staff for the Home or Reserve Army that had formerly been under the command of Col. Gen. Friedrich Fromm, an early ally of Dr. Speer.

The SS even had its own rocket-testing site at Blizna in German-occupied Poland, located between Lvov and Krakow, also run by Kammler, now Speer's vigorous rival; the Red Army took Blizna, however, in July 1944. Of Kammler, Gitta Sereny wrote that he was, "In looks and ruthlessness almost a reincarnation of Reinhard Heydrich."[15] Speer recalled: "I knew him when he was running the construction division in the Ministry of Air Transport: an inconspicuous, sociable, and very hardworking official. Nobody would have dreamed that some day he would be one of Himmler's most brutal and most ruthless henchmen."[16] He disappeared at the end of the war, and is rumored to have worked secretly for the US or Russia.

"The ruthless builder of the Auschwitz gas chambers and the *Mittelwerk* (at Dora), SS Major General Hans Kammler became the dominant personality in the Army rocket program after the assassination attempt against Hitler in July 1944," wrote Speer biographer Gitta Sereny. (CER)

thrill provided by the huge missile in flight, nor seen a place where one had hit the ground, so now we had to give him a convincing demonstration.[17]

They stayed at Rastenburg's Army Guest House, Hunter's Height, and went to see Hitler, after 5p.m. "Suddenly, the door opened, and we heard someone call out, 'The Führer!' Hitler appeared in the company of Keitel, Jodl, [Gen. Walter] Bühle, Speer, and their personal aides. No visitors were allowed. I was shocked at the change in Hitler. A voluminous black cape covered his bowed, hunched shoulders and bent back. He looked a tired man."[18] Hitler had lost the Battle of Stalingrad the previous February, and was very soon to lose the battle of Kursk.

The filmstrip concluded with the words "We made it after all!" flashed across the whole screen. Von Braun concluded, "Hitler was visibly moved and agitated... He showed real interest." Major General Dr. Dornberger recalled:

Hitler jumped up and crossed over to the table on which we had arranged our little show of models... Hitler came over and shook my hand. He said in a whisper, "I thank you! Why was it I could not believe in the success of your work? If we had had these rockets in 1939, we should never have had this war. Europe and the world will be too small from now on to contain a war. With such weapons, humanity will be unable to endure it."

A prototype V-2 rocket in test colors of black and white about to be fired. (CER)

He turned again to the model of the bunker. We had to take it all to pieces again and explain everything a second time. We had to show him how the rockets could be brought up, stored, tested, and got ready for launching; then how, after being set up, they could be moved through the narrow sliding doors into the open, one minute before firing, with their interior gyroscopes running, and then launched almost immediately from their vertical position on the table...

Hitler interrupted me impulsively and called Speer over to tell him that it was these same large-scale bunkers on the Channel coast that had proved so useful for submarines.

He wanted to have not one but two, or if possible three, bunkers put up for us. Motorized rocket batteries, in his opinion, would soon be spotted by enemy reconnaissance and engaged. The future was to prove him wrong. Although I knew that Hitler could not stand contradiction, I did not hesitate to contest his view... My arguments were in vain.

Bunkers were Hitler's favorite buildings, and he would not drop the idea. Speer received orders to have the roof of the bunker built to a thickness of 23ft. Hitler added this explanation of the plan: "These shelters must lure the enemy airmen like flies to a honey pot. Every bomb that drops on them will mean one less for Germany!"[19]

Hitler also wanted even larger rockets built:

...to raise the load of explosives to 10 tons and the monthly deliveries to 2,000... It would take at least four or five years to develop such a gigantic rocket. "What about quantities?" Hitler interposed impatiently. "Not possible either," I explained. "We haven't enough alcohol."

A strange, fanatical light flared up in Hitler's eyes. I feared he was going to break out in one of his mad rages. "But what I want is annihilation—annihilating effect!" ... Speer, Keitel, Jodl, Bühle, and the others stood silent and apart, watching me closely. I replied briefly,

As one V-2 lifts off from Stand #7 at Peenemünde in June 1943, a second missile (left) is raised into the firing position and third (foreground) sits on its *Meilerwagen* mobile launcher. (CER)

"No one can get more out of a ton of explosive than it is capable of giving..."

"Please discourage the propaganda that is starting about the decisive effect these 'all-annihilating wonder weapons' are going to have on the war," I urged. "It can lead to nothing but disappointment for the populace..."

"When we started our development work, we were not thinking of an all-annihilating effect. We—" Hitler swung around in a rage and shouted at me, "*You*! No, *you* didn't think of it, I know, but *I* did!" In face of this outburst, I decided to keep silent. Keitel hastened to change the subject by stressing the need for more air raid defenses at Peenemünde. Antiaircraft guns were granted forthwith. The tension relaxed.[20]

Hitler promised that the V-2 program would now be top priority, then promoted von Braun to the rank of professor on the spot. Walking toward the doorway to leave the room, the Führer then turned and came back to Dr. Dornberger: "I have had to apologize to only two men in my life. The first is Field Marshal von Brauchitsch. I did not listen to him when he told me again and again how important your research was. The second man is yourself. I never believed that your work would be successful."[21]

The object of this fulsome compliment found himself with mixed feelings:

I saw myself exposed to the dangerously dynamic personality of this unpredictable man, with his possibly exaggerated hopes... The catch word was "wonder weapons"... Our A-4—the long-range missile with the simple workshop name 'Aggregate #4'—had been turned into a V-2, standing for "Vengeance Weapon #2!"...

It was by no means a "wonder weapon." ... an exaggeration which did not correspond with the facts. By the middle of 1943, the military situation had long ceased to be such that by launching 900 V-2s a month, each loaded with a ton of explosive over ranges of 160 miles, one could end the Second World War. I was haunted by forebodings.[22]

That July 1943—from the new Armaments Ministry headquarters at the Zoo Bunker in Berlin—Speer Ministry Central Office head Karl Otto Saur announced the new "Degenkolb Program" for producing 2,000 V-2s monthly as of the following December.

On the "Black Day" of Tuesday, August 18, 1942, however, a massive RAF raid on Peenemünde—Operation *Hydra*—wrecked the Saur-Degenkolb Program, as Major General Dornberger recalled in 1954: "600 four-engine bombers are said to have taken part in the raid with 1,500 tons of bombs dropped; 47 bombers were shot down by antiaircraft guns and night fighters."[23]

RAF Air Marshal Sir Arthur Harris—now to become one of Speer's most tenacious Allied foes—had sent 596 heavy bombers and 4,000 crewmen against Peenemünde.[24] The result was 732 dead and 800 injured. Peenemünde was finished as a production site because the RAF could always hit it again.

Meanwhile British Intelligence—which had learned of Nazi Germany's rocketry program in 1939—acquired aerial photos of Peenemünde's V-1 ski launching ramps, and thus in December 1943 Churchill launched Operation *Crossbow* to destroy them.

On June 13, 1944, the reality of the V-2's existence became known when #4089 exploded in an airburst over neutral Kalmar near Malmö, Sweden. The Swedes turned over the remains to the Allies, fearing a Nazi invasion if the Germans won the war.

Writing in his own memoirs, Gen. Dwight D. Eisenhower asserted that he always feared that Nazi rockets would be used against his embarkation points for Operation *Overlord*, the Allied invasion of northwestern Europe, rather than London.[25]

Speer wrote in his Spandau diary that "in the summer of 1944, Field Marshal Keitel, Chief of Staff Col. Gen. Kurt Zeitzler, armaments manufacturer Hermann Röchling, Porsche, and I sat with Hitler to discuss the crisis caused by the knocking out of the German fuel production plants. 'Soon we can begin the attacks on London with the V-1 and V-2,' Hitler declared exultantly. 'A V-3 and V-4 will follow, until London is one vast heap of ruins. The British will suffer! They'll find out what retaliation is! Terror will be smashed by terror. I learned that principle in the street battles between the SA and the Red Front.'"[26]

Following the successful D-Day invasion, on November 11, 1944, US Army Gen. George C. Marshall and US Army Air Force Gen. Henry A. "Hap" Arnold authorized a massive aerial strike against 150 known and suspected V-1 ski launching ramp sites in Nazi-occupied France.

In April 1943, Reichsführer-SS Heinrich Himmler had made his own initial Peenemünde inspection tour, squired around by Dr. Dornberger. Himmler told Dornberger: "Once the Führer has decided to give your project his support, your work ceases to be the concern of the Army Weapons Department, and becomes the concern of the German people. I am here to protect you against sabotage and treason."[27]

In effect, Himmler had made his first move to acquire the V-weapons as part of his planned postwar SS state. Dornberger countered that, since Peenemünde was an Army facility, internal security was rightfully headed by Colonel General Fromm, but as a compromise, accepted SS external security surrounding the station. Himmler concurred, and immediately appointed the police commissioner for nearby Stettin on the Baltic coast, SS-Obergruppenführer Emil Mazuw, who was with him at the meeting.

A week later, Dr. Dornberger was identified as a "brake" on the development of missiles by SS Captain Engel, head of the newly established SS Rocket Research Station Grossendorf, near Danzig; already therefore, Himmler was setting up his *own* SS shadow rocketry

"An A-4B—forerunner of the A-9—prior to being fired from Test Stand #10 in January 1945," according to the Smithsonian National Air and Space Museum. (CER)

program to mimic the actual one. Seeking to buttress his internal political base, Dr. Dornberger immediately drafted and published *Rocket Development: The Achievement of the Army Weapons Department, 1930–43.*

Reichsführer-SS Himmler made his second visit to Peenemünde on June 29, 1943: "Himmler arrived unaccompanied, driving his own private little armored car ... Himmler possessed the rare gift of attentive listening... He grasped unerringly what the technicians told him... A man without nerves."[28]

As Dr. Speer recorded, Himmler succeeded in taking control of rocket production away from him via a Führer decree of August 20, 1943, following the RAF bombing of Peenemünde.

In September 1943, Himmler made his move, appointing his subordinate—SS Brigade Leader Dr. Hans Kammler, head of the building branch of the SS Main Office—as *his* man for all things Peenemünde, to take charge of the A-4 special committee within Professor Speer's own Ministry of Armaments, thus simultaneously infiltrating that as well.

Under the new rocketry chain of command—Himmler, Speer, SS-Obergruppenführers Oswald Pohl, and Dr. Hans Kammler—5,000 A-4/V-2 rockets would be manufactured by January 1945. Thus, German rocketry production was now split three ways: under Home Army Colonel General Fromm, under Munitions officials Dr. Speer and Saur, and also subordinated to Himmler and Kammler. The V-1 was a Luftwaffe rocket under Göring, while the V-2 was a stool with three legs: Army, SS, and Speer's Munitions Ministry.

General Dornberger's first meeting with Dr. Kammler was in Berlin on September 6, 1943: "He was simply incapable of listening. His one desire was to command... He made his decisions without due consideration. He rarely ever conceded any point ... arrogant, brutal, overbearing, intolerably haughty to those below him."[29]

This, then, was the man who was finally to dominate German rocketry production until the very end of the war, and who thus bedeviled not only Dr. Dornberger and Dr. von Braun, but also Professor Speer, who wrote of him:

I knew him when he was running the construction division in the Ministry of Air Transport. Nobody would have dreamed that someday he would be one of Himmler's most brutal and most ruthless henchmen. It was simply inconceivable, and yet that was what happened...

Kammler—who made an extremely fresh, energetic, and ruthless impression, comparable to Heydrich's—had begun by taking over relatively small tasks of the A-4 production within the overall armaments area. Then he had assumed responsibility for rocket launchings, which was actually a military task. Finally, he obtained the production of all special weapons on the basis of rockets; and at the close of the war, he also received responsibility for manufacturing all jet airplanes.

DOCTOR WERNHER VON BRAUN (1912-77)

The man the Führer depended upon to realize his dream, of the destruction of the enemy's cities in fire, was the heir to a German baronic title: Wernher von Braun, born on March 23, 1912 at Wirsitz, Germany (now Wyrzysk, Poland), the son of Baron Magnus von Braun and his science-loving wife.

As a child, he showed a precocious interest in rocketry, and by the time he was 21, he had outlined the design for a moon rocket. In college, he became an assistant to German rocket pioneer Hermann Oberth. After he graduated in 1932, he was employed by the German Army and joined Dornberger's Experimental Station. Though regarded by many as a genius, others have stressed that his strongest achievements were as manager, salesman and self-promoter—not dissimilar to Speer.[30]

On May 15, 1937, at the age of 25, von Braun was named technical director of the Peenemünde Rocket Research Facility on the Baltic coast. In March 1944, von Braun was arrested by the SS, officially for giving peacetime projects precedence over war production, but he was released through the intercession of Albert Speer."[31]

After surrendering to US Army troops, von Braun became part of Operation *Paperclip*, living and working in America, eventually becoming the patriarch of American rocketry. Working first for the US Army, then NASA, he was part of the team that developed the Redstone missile, and launched the Pioneer 3 and 4 satellites into orbit. In 1956

Stated the Smithsonian National Air and Space Museum at Washington, DC, "The legacy of the V-2s in America: after their transfer to the Redstone Arsenal in Huntsville, AL, von Braun and his staff developed an extensive family of missiles based upon the V-2... From left to right: General Electric's Hercules missile (a copy of the German Wasserfall antiaircraft missile), a V-2 (with experimental paint configuration), a Juno missile, a Redstone missile, a Saturn I launch vehicle, and a Jupiter C missile. The Jupiter C holds a mock-up of Explorer I." (NASA photo)

Left: Von Braun (right) in the US after the war, seen here with US Army Ordnance Colonel Holger Toftoy at White Sands, NM. According to the National Air and Space Museum in Washington, DC: "The US Army entered the ballistic missile race among US military services partly by bringing captured V-2 missile parts——along with project personnel——to White Sands Proving Ground, NM, for evaluation and testing…von Braun and his team advised General Electric and Army Ordnance personnel in the reassembly, testing, and firing of the missiles. The first firings used all-German components. With time, American-made components were substituted to gain working experience with refining the basic missile design." (Smithsonian Air and Space Museum, Washington, DC.)

von Braun was appointed Director of Development of the Operations Division, Army Ballistic Missile Agency. In 1960 he became director of the George C. Marshall Space Flight Center at Huntsville, AL. In the same year, a biopic of his life entitled *I Aim at the Stars*, was released. In 1969, von Braun fulfilled his dreams with the launch of the Saturn V rocket that took American astronauts to the moon. Von Braun left NASA in 1972 to become an executive of Fairchild Industries. From 1975 he was President of the National Space Institute. He died in 1977 from cancer. After his death, President Jimmy Carter said: "Wernher von Braun's name was inextricably linked to our exploration of space. Not just the people of our nation, but all the people of the world have profited from his work."[32]

Below: During the presidencies of Eisenhower, Kennedy, Johnson, Nixon, Ford, and Carter, von Braun was at the very epicenter of the American missile, rocket, and space programs. Here, on November 16, 1963—just a week before the Dallas assassination—he (center) and President John F. Kennedy (right) are taking a tour of NASA's Cape Canaveral missile launch site in Florida. At far left is Robert Seamans, NASA's deputy administrator. (NASA photo, JFK Library, Boston, MA)

At the last minute, Hitler put Kammler in charge of all air armaments. Thus—just a few weeks before the end of the war—he had become commissioner general for all-important weapons. Himmler's goal was achieved, but there was no more armaments industry... The goal that Himmler had always striven for was realized: that the technicians and engineers of armaments should run the SS works under the control of the SS."[33]

Dr. Kammler was also in charge of the SS slave labor component of the overall rocketry program. According to Speer:

Kammler's "Special Staff" was headed by "Dr. Kammler's Construction Bureau," with its main office in Berlin. This office commanded "SS Special Inspections," which were distributed throughout the Reich and, in turn, headed the regional "SS Command Staffs." The files reveal that the SS remained in charge of employing and controlling the prisoners used under Kammler...

Under Kammler—a relentless but capable robot—there was a new development in the spring of 1944: ultimately it threw hundreds of thousands of prisoners into the production process under the harshest conditions...[34]

It it is not really known what happened to SS-Obergruppenführer Dr. Hans Kammler at the end of the war. The rumor was that he was shot by his own SS aide in the last days of the war, but it is possible that he escaped Allied justice altogether, or possibly defected to the Russians, to produce their rocketry program, or to America.

In January 1944, a great setback occurred when V-2s began exploding during the early part of their aerial flights. That June, the SS established its own testing ground at Heidelager, near Blizna, Poland, with an impact area in the Pripet Marshes, 200 miles to the northeast: in effect, an Army facility on SS grounds.

Following the failed Army bomb plot to assassinate Hitler in July 1944, the Führer appointed Reichsführer-SS Himmler to replace plotter Col. Gen. Fritz Fromm as head of the Home Army; the previous October, Himmler had also taken over the post of his nominal superior—Dr. Wilhelm Frick—as German Reich Minister of the Interior. On August 8, 1944, Himmler appointed SS Lieutenant General Kammler as head of the entire V-2 production program. Dornberger's assessment was that "An uninformed layman took the leadership," with all the various powers that he himself had sought for so long, but had failed to achieve.[35]

That September, "The V-2 was put into military operation," according to Dr. Dornberger.[36] On September 8, 1944, the first V-2 was launched from Holland on London by Mobile Artillery Section 485. "From September onward, rockets were transported to the front. In September, 350 were delivered, in October 500, then between 600–900 every month."[37]

A cutaway diagram of the German Wasserfall (Waterfall) antiaircraft rocket. Speer believed that the USAAF's Flying Fortresses were doing far more damage than all the Third Reich's V-weapons put together, and wanted mass production of the antiaircraft rockets instead. "Project Waterfall could have gone into production even earlier had Peenemünde's full capacity been used in time. As late as January 1, 1945, there were 2,210 scientists and engineers working on the long-range rockets A-4 and A-9, whereas only 220 had been assigned to Waterfall, and 135 to another antiaircraft rocket project, Typhoon." Reich Marshal Hermann Göring saw a Wasserfall launch on October 30, 1944 and exclaimed, "That's terrific! We must have that at the first Party Rally after the war!" (LC)

Whatever else can be said, the team of Himmler and Dr. Kammler had succeeded where that of Dr. Speer, Dr. Dornberger, and Dr. von Braun had failed. Hitler had, at least in part, achieved his goal, as the rockets actually arrived at the front and went into action. One must grant him that. The SS had succeeded where the Army had dithered.

Late in June 1944, construction of Site S-3 began in Thuringia's Harz Mountains, but it was never completed. Hitler apparently authorized this new building project, although Speer was unaware of it. It could conceivably have been a secret SS FHQ, as Himmler's chief of construction—Hans Kammler—was in charge, rather than Speer's own OT. Supposedly, the Reichsführer-SS planned to present Hitler with FHQ S-3 as a birthday gift in 1945, but it is also possible that Himmler planned to use it himself as Germany's first SS Führer had he taken over in a coup. To this day, no one knows for sure the real purpose of S-3, and it remains one of the unsolved mysteries of the war.[38]

This also may be one of the many reasons why the SS-run underground rocketry factory named Dora—or Mittelwerk (Central Works) Nordhausen—was built in the Thuringian Harz Mountains. I think it is quite possible that the SS FHQ was built to physically command the SS rocket program and thus continue the war. The factory

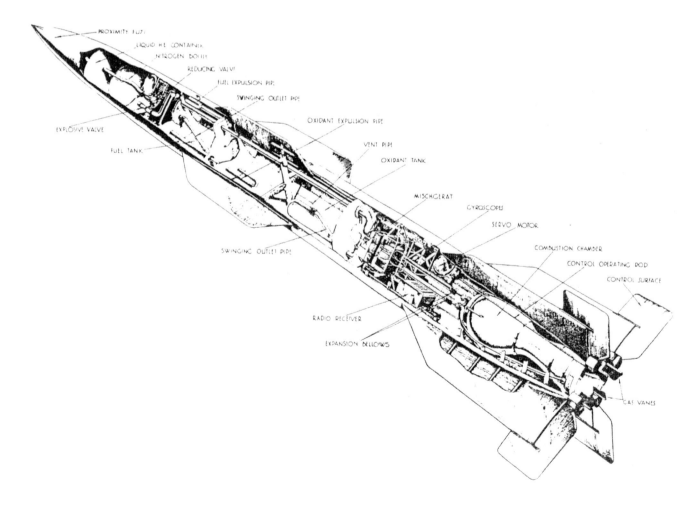

was officially ordered created on August 22, 1943 by a Führer Order issued at FHQ Wolf's Lair at a meeting of Hitler, Dr. Speer, and Himmler.

With the introduction of the SS slave labor program to be run by Dr. Speer's Ministry of Armaments managers and engineers at Hitler's own sanction, the new Central Works was inaugurated at Dora, and thus—in Speer's own words—"Himmler had put a foot in our door, and we ourselves had helped him do it."[39]

On December 10, 1943, Speer inspected Dora for the first time:

> In a lonely valley in the Harz Mountains a widely ramified system of caves had been established before the war for the storage of vital military chemicals... I inspected the extensive underground installations where the V-2 was to be produced. In enormous long halls, prisoners were busy setting up machinery and shifting plumbing. Expressionlessly, they looked right through us, mechanically removing their prisoners' caps of blue twill until our group had passed them.
>
> The conditions for these prisoners were in fact barbarous... The sanitary conditions were inadequate, disease was rampant; the prisoners were quartered right there in the damp caves, and as a result, the mortality among them was extraordinary high.[40]

Above and below: A German Wasserfall rocket (C2-F/45) in 1945. It had a conventional warhead and weighed about 3.5 tons when fully loaded. Noted one source: "This joint Luftwaffe-Army antiaircraft missile became Peenemünde's second major project in the last two years of the war." Writing in 1969, Dr. Speer still believed that use of the Me 262 as jet fighters and the Wasserfall rockets could have turned back the Allied air offensive in the spring of 1944, before the D-Day invasion. (Photos by Beth A. Goodrich, US Army Ordnance Corps Public Affairs Office, Aberdeen Proving Ground, MD)

Gitta Sereny termed Dora "the most hellish of any" of the Nazi labor camps.[41] Its workers were drawn from the nearby SS Buchenwald concentration camp. The various spheres of Dora responsibilities were broken down thus: Dr. Dornberger and Dr. von Braun headed the purely technical sector, Professor Speer the financing, and Dr. Kammler's SS the construction.

Dr. Speer himself called it "The worst place I have ever seen... The prisoners lived in the caves with the rockets" that they themselves built. "I saw dead men... They couldn't hide the truth."[42]

He claimed that, as a result, in March 1944, he urged the SS to improve living, working, and housing conditions. To an extent, this was done—in order to get more production of work from the prisoners. "I loved machines more than people," he admitted.[43] New barracks were built at Dora and the 31 sub-camps that surrounded it; 60,000 men were sent to Dora, and fully half of them died there. "I was appalled," Speer told Sereny decades later.[44] The Ministry of Armaments' man in charge under his direction was Georg Rickhey.

In the 1980s, Arthur Rudolph, then retired from his work with NASA, was investigated for war crimes by the Office of Special Investigations. The investigation centered on his role in the treatment of laborers at Dora. Eventually he signed an agreement whereby he left the US, and renounced his US citizenship. He returned to Germany where an investigation took place before he was allowed German citizenship. In my opinion, Rudolph was being used as a scapegoat for the dead von Braun, and their joint direction of the Nazi underground rocket factory at Nordhausen, Germany.[45]

In one other area did the SS prove Professor Speer wrong as well: in December 1944—even after the collapse of the rest of the German armaments industry—Degenkolb managed to produce no fewer than 613 A-4/V-2 rockets.

The Red Army began closing in on the SS rocket testing ground at Heidelager in Poland by December 1944. That same month, Peenemünde—defended by Himmler's Volkssturm (People's Army)—was also taken by the Soviets. Dornberger recalled the unfolding of events:

> On January 27, 1945, the whole "Working Staff Dornberger" met in Berlin for the first time... The situation and prospects were horribly depressing... All the firms concerned were to be evacuated to the Nordhausen-Bleicherode area in Thuringia...
>
> The move began early in February. My own staffs also moved from Schwedt on the Oder River to the southern slopes of the Harz Mountains near Bad Sachsa. All other firms in our program, as well as the Service departments—whether Army or Air Force—were evacuated to the same area.
>
> ...Even if the High Command had succeeded in holding the Americans and British on the Rhine and the Russians on the Vistula, bombing alone would have delayed until the spring of 1946 delivery of any of the rockets in sufficient quantity to protect vital objectives in the front line and at home and give any substantial relief in the air war...[46]

Meanwhile, the Dutch V-1 and V-2 launching sites were overrun by the forces of British Field Marshal Bernard Law Montgomery. Some 6,000 V-2s were built, approximately half of which were launched. Of the 1,054 launched against London, 517 hit the city. The Belgian post of Antwerp fared even worse, receiving no fewer than 1,265 hits. On April 3, 1945, Dr. Kammler gave his last A-4 production orders, but they weren't carried out. The plan to fit A-4 rockets on U-boats to attack American ports from offshore never happened, and the projected gigantic A-10 rockets to attack Washington, DC and New York were never built.

Also on April 3, 1945, SS-Obergruppenführer Kammler gave orders to evacuate the staffs of 450 rocket scientists and engineers to Oberrammergau in west Germany in the Allgäu Mountains, and these were obeyed. Thus, all of them were taken prisoner on May 2, 1945 by the advancing US Seventh Army of Gen. Alexander M. Patch near Hindelang, just as Dr. Dornberger and Dr. von Braun wanted. Nordhausen was taken by the US Army on April 10, 1945, a week after Dr. Kammler's last rocket orders had been given.

On July 19, 1945, the American Operation *Overcast/Paperclip* was launched to relocate German rocket scientists and engineers to the United States, where they would be employed to jump-start the Americans' own postwar missile race against the Free World's new enemy: the USSR of Josef Stalin.

CHAPTER EIGHT

SPEER AND THE GERMAN ATOMIC BOMB, 1942–45

Schirach has been ill for several days, and so I have been taking longer walks with Hess... Today I told him about my occasional unauthorized acts as armaments minister ... of the switch of our atomic research to a uranium-powered motor because Heisenberg could not promise to complete the bomb in less than three to five years.

That I had acted without authority stirred Hess to excitement. 'You mean to say that you didn't send up a query about the atom bomb?' he interjected, dismayed. "No, I decided that on my own. At the end, it was no longer possible to talk with Hitler."

In the style of an official Party reprimand, Hess reproved me: "Herr Speer, that is an impossible proceeding. I really must point that out. The Führer had to be informed in order to be able to make the decision himself." After a pause, he added, "What things came to after I left!"

Albert Speer, *Spandau: the secret diaries*, December 2, 1962

The German Jewish physicist Dr. Albert Einstein (1879–1955), who left Germany after 1933, and offered his services to the United States instead. In 1939, he wrote President Franklin D. Roosevelt a famous letter in which he warned FDR that the Nazis were capable of developing an atomic bomb. The top-secret American Manhattan Project was created as a result. (LC)

At the height of the Blitz on Britain in 1940, Nazi scientists were fully aware that American scientists were dedicated to winning the race to develop the atomic (or uranium) bomb. Their information seems to have come from colleagues in America, an information leakage which may not have stopped with Pearl Harbor. However, the quality of the information was "third rate," and they had little idea how far Allied research had come, for in 1945 Heisenberg and other leading scientists were "'completely staggered' by the news that the Americans had in fact dropped the weapon on Japan on August 6, 1945."[1]

Their reaction to the first use of the atomic bomb, and subsequent discussions were secretly recorded by British Intelligence when they were in England for debriefing, and their incarceration was "timed to coincide with the dropping of atomic bombs on Hiroshima and Nagasaki."[2] The group speculated how the course of the war might have been altered had Germany had the bomb: "They speculated that such a weapon in Germany's hands would inevitably have led to the destruction of London and to the development of such science fiction weaponry as 'long-distance aircraft with uranium engines;' and for the historians they left something of a mystery: Did these German scientists purposefully stall the Nazi nuclear program?"[3] I pose this question: did Speer also?

When news of the bomb was given to the scientists, including Otto Hahn, who had actually discovered uranium fission in 1939, the bugs

The German nuclear effort in 1940, showing the locations of the main research and industrial work on atomic energy after the German invasion of France. (LC)

recorded "what became an orgy of self pity, apparently conflicted emotions, and face-saving protestations."[4] A leading researcher, Walter Gerlach said:

> "When we go back to Germany, we will have a terrible time. We will be looked upon as the ones who sabotaged everything. We won't remain alive long out there."

Hahn asked, "Are you upset because we didn't make the bomb? I thank God on my bended knees that we didn't make a uranium bomb.

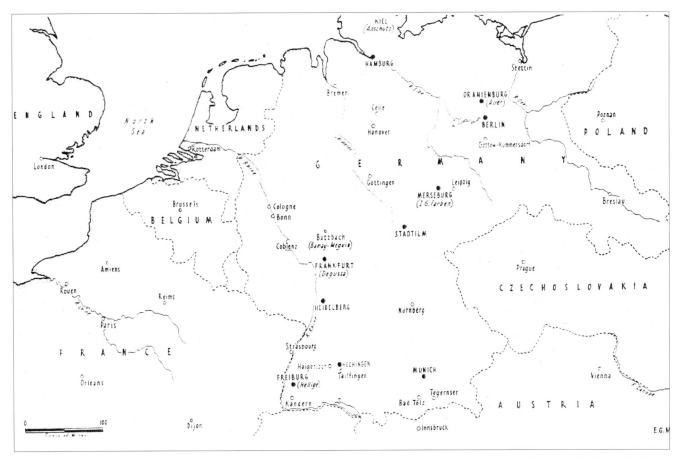

The German nuclear effort in 1944, showing the locations of the main research and industrial work on atomic energy in the last months of the war. (LC)

Are you upset because the Americans did?" When Gerlach replied, "Yes," Hahn responded, "Surely you are not in favor of such an inhuman weapon?"

Said Gerlach, "No, we never worked on a bomb, but I didn't believe that it would go so quickly. But I did think that we should exploit the energy for the future."[5]

Later in the day, at the dinner table, they continued discussing what might have been:

Carl-Friedrich von Weizsäcker discussed what the Nazi scientists might have done with a nuclear program—and when they might have done it. "If we had wanted to make the bomb," said von Weizsäcker, "we should've concentrated more on the separation of the isotopes and less on heavy water…"

Weizsäcker continued, "If we had started the business soon enough, we could have got somewhere. If the Americans were able to complete it in the summer of 1945, we might have had the luck to explode it in the winter of 1944–45."

Karl Wirtz responded with a rather chilling thought: "The result would have been that we would have obliterated London, but we still would not have conquered the world. What would have happened to us then?"

Von Weizsäcker then suggested that the scientists feared their own success, that conscience had gotten the better of them. "We didn't succeed because we didn't want to succeed," said von Weizsäcker. "If we had put as much energy into it as the Americans—and had wanted it much as they did—we could have done it," and he added, "It would have been a much greater tragedy for the world if Germany had had the uranium bomb. Just imagine if we had destroyed London with the bomb, it would not have ended the war, and when the war had ended, it is still doubtful if it would have been a good thing."[6]

Crookland noted that some of Heisenberg's words could be taken as evidence of sabotage:

> Later, Heisenberg raised a technical question about the world's first nuclear devices, hinting that he might have played a small part in sabotaging the German program, and he still doubted whether an actual nuclear breakthrough had been made: "I still don't believe a word about the bomb, but I may be wrong." Said Heisenberg, "I consider it perfectly possible that they have 10 tons of pure uranium 235."
>
> Hahn picked up immediately on this statement: "I thought that one needed very little U-235? You used to tell me that one needed 50 kilograms of U-235 to do anything. Now you say one needs 10 tons!" Heisenberg responded, "I wouldn't like to commit myself for the moment... I would say that, at the rate we were going, we would not have succeeded during the war."
>
> ... Whether Heisenberg's miscalculation about the amount of raw materials needed was a genuine mistake or an attempt to mislead—and thus deter Hitler from pressing for a bomb—is open to a most intriguing sort of speculation.[7]

It should be here noted that the American scientists also faced these very same moral dilemmas and qualms—then produced the atomic bomb anyway. We turn now to Speer's involvement in Germany's development of the uranium bomb. Writing in 1970, he recalled:

> I met regularly for lunch with Gen. Friedrich Fromm ... at Horcher's Restaurant... At the end of April 1942, he remarked that our only chance of winning the war lay in developing a weapon with totally new effects. He said he had contacts with a group of scientists who were on the track of a weapon that would annihilate whole cities, and perhaps throw the island of England out of the fight. Fromm proposed that we pay a joint visit to these men. It seemed to him important, he said, to at least have spoken with them.
>
> Dr. Albert Vögler—head of the largest German steel company and President of the Kaiser Wilhelm Institute—also called my attention at this time to the neglected field of nuclear research. He complained of

the inadequate support fundamental research was receiving from the Ministry of Education and Science [headed by Minister Bernhard Rust], which naturally did not have much influence during wartime.

On May 6, 1942, I discussed this situation with Hitler and proposed that Göring be placed at the head of the Reich Research Council—thus emphasizing its importance... On June 9, 1942, Göring was appointed to this post.

The three military representatives of armaments production—Milch, Fromm, and Witzell—met with me at Harnack House, the Berlin Center of the Kaiser Wilhelm Institute, to be briefed on the subject of German atomic research... The subsequent Nobel Prize winners Otto Hahn and Werner Heisenberg were present.

After a few demonstration lectures on the matter as a while, Heisenberg reported on atom smashing and the development of a uranium machine and the cyclotron... America probably had a head start in the matter, whereas Germany had been in the forefront ... only a few years ago. In view of the revolutionary possibilities of nuclear fission, dominance in this field was fraught with enormous consequences.

...I asked Heisenberg how nuclear physics could be applied to the manufacture of atom bombs... He declared ... that the scientific solution had already been found and that, theoretically, nothing stood in the way of building such a bomb, but the technical prerequisites for production would take years to develop, two years at the earliest.

Difficulties were compounded, Heisenberg explained, by the fact that Europe possessed only one cyclotron, and that of minimal capacity ... located in Paris, and because of the need for secrecy, could not be used to full advantage. I proposed that—with the powers at my disposal as Minister of Armaments—we build cyclotrons as large as, or larger than, those in the United States, but Heisenberg said that because we lacked experience, we would have to begin by building only a relatively small type.[8]

Despite the fact that Dr. Speer alleged that he offered millions of Reich Marks and men to support an upgrading of the overall German atomic bomb effort, "I had been given the impression that the atom bomb could no longer have any bearing on the course of the war." On June 23, 1942, Dr. Speer reported to Hitler "only very briefly" on the matter.

As he recalled, however, Hitler had several other sources of information from different members within both his inner circle and the official members of the regime:

Hitler had sometimes spoken to me about the possibility of an atom bomb, but the idea quite obviously strained his intellectual capacity. He was also unable to grasp the revolutionary nature of nuclear physics. In the 2,200 recorded points of my conferences with Hitler, nuclear fission comes up only once, and then is mentioned with extreme brevity.

Hitler did sometimes comment on its prospects, but what I told him of my conference with the physicists confirmed his view that there was not much profit in the matter.

Actually, Professor Heisenberg had not given any final answer to my question whether a successful nuclear fission could be kept under control with absolute certainty, or might continue as a chain reaction. Hitler was plainly not delighted with the possibility that the earth under his rule might be transformed into a glowing star.

Occasionally, however, he joked that the scientists in their worldly urge to lay bare all secrets under heaven might some day set the globe on fire...[9]

The feared Ju 87 Stuka dive-bomber designed and promoted by Göring's technical office chief, Col. Gen. Ernst Udet, an ace in the Great War with 62 kills. Successful during 1939–40, it was found to be obsolete during the Battle of Britain, but continued to see active combat service on the Russian Front during 1941–45 nonetheless, sinking the Russian Navy battleship *Marat* and destroying thousands of T-34 tanks. (CER)

Ironically, that is exactly the possibility that American President Harry S. Truman's scientists laid out for him in 1945 if the United States exploded the atomic bomb; he decided to take the risk, a huge gamble by any standard. Speer commented that he was quite sure Hitler would not have hesitated to use atomic bombs against England.

Speer dropped the project to develop an atomic bomb by autumn 1942, following discussion with the nuclear physicists which indicated they were several years away from a usable weapon. Instead, he authorized the development of an energy-producing uranium motor for propelling machinery.

Going back to the Speer–Hess dialogue at the beginning of this chapter, I must again state that I agree with Hess: had Hitler been given the decision to make, he would have stepped up the program with all available resources immediately and decisively. Speer continued:

At Heidelberg in the summer of 1944, I was shown our first cyclotron splitting an atomic nucleus. To my questions, Professor Walther Bothe explained that this cyclotron would be useful for medical and biological research. I had to rest content with that.

In the summer of 1943, wolframite imports from Portugal were cut off, which created a critical situation for the production of solid-core ammunition. I thereupon ordered the use of uranium cores for this type of ammunition. My release of our uranium stocks of about 1,200 metric tons showed that we no longer had any thought of producing atom bombs.[10]

One wonders whether Allied Intelligence were aware of this? In 1970 Speer admitted that: "Perhaps it would have proved possible to have the atom bomb ready for employment in 1945, but it would have meant mobilizing all our technical and financial resources to that end, as well as our scientific talent. It would have meant giving up all other projects, such as the development of the rocket weapons. From this point of view, too, Peenemünde was not only our biggest, but our most misguided project."[11]

Hitler was also prejudiced about the Jewish influence on American atomic development, referring to "Jewish physics," an attitude shared by Rosenberg. But Speer argued:

> …even if Hitler had not had this prejudice against nuclear research, and even if the state of our fundamental research in June 1942 could have freed several billion instead of several million Marks for the production of atom bombs, it would have been impossible—given the strain on our economic resources—to have provided the materials, priorities, and technical workers corresponding to such an investment.
>
> For it was not only superior productive capacity that allowed the United States to undertake this gigantic project; the increasing air raids had long since created an armaments emergency in Germany, which ruled out any such ambitious enterprise. At best—with extreme concentration of all our resources—we could have had a German atom bomb by 1947, but certainly we could not beat the Americans, whose bomb was ready by August 1945, and on the other hand, the consumption of our latest reserves of chromium ore would have ended the war by January 1, 1946, at the very latest.
>
> From 1937–40, the Army spent 550 million Marks on the development of a large rocket, but success was out of the question, for Hitler's principle of scattering responsibility meant that even scientific research teams were divided and often at odds with one another. According to the *Office Journal* for August 17, 1944, not only the three branches of the armed forces, but also other organizations, the SS, the postal system, and such, had separate research facilities. In the United States, on the other hand, all the atomic physicists—to take an example—were in one organization.[12]

While that is true, President Roosevelt was also notorious for assigning the same job to several different people, and then taking the best results from the most successful.

As Gitta Sereny pointed out in her biography of Speer, he had a more positive spin on atom bombs in his initial 1953 Spandau draft of his memoirs than the published version, probably changed once he had experienced the attitudes of young people in the late 1960s.[13] In his 1953 draft, he wrote:

> Thanks to the insane hatred of the leadership, we allowed ourselves to lose a weapon of decisive importance. If, instead of backing the—in the final analysis—ineffective rockets with hundreds of millions, we had devoted them to supporting atom research from the start, it would have been more useful for the war...[14]

She then stated that Speer not only failed to accurately inform Hitler, but also his nominal superior Göring under the Four Year Plan, and for the exact same reason: the Reich Marshal—like his Führer—would have immediately pushed for a speeded-up program of development.

She also revealed that Professor Heisenberg—at the September 1941 Copenhagen meeting with the Danish scientist Niels Bohr—had tried to get the latter to join the Nazi team, since the Reich was going to win the war, and develop the atomic bomb before the Allies. After the war, Heisenberg claimed that Bohr had misunderstood him, and that he'd really approached him about not producing any atomic bomb, as discussed in Thomas Powers' book, *Heisenberg's War*. Her conclusion was: "Powers accepted some of Heisenberg's claims, which are now shown by Speer's Spandau account to be false." Sereny added:

> Speer's various versions of events also differ ... 15 years earlier, however, in the "Spandau draft," on July 3, 1953, still in ignorance of what German physicists were claiming, he wrote that *it was only around the time of the Normandy landings, in June 1944* that the atom scientists came to see him to admit there was no hope of an atom bomb for several years.[15]

Only later would he claim that they said this to him in 1942. Sereny continues:

> This timing is borne out by his discussion with Hess, on December 2, 1962, which he reports in *The Secret Diaries*, written five years after *Inside the Third Reich*. Hess was astonished, he writes there, that Speer had on his own responsibility switched the scientists' efforts toward a motor instead of a bomb.
>
> "You mean to say that you didn't send up a query about the bomb?" he interjected, dismayed. Speer said no, he decided on his own—"At the end, it was no longer possible to talk with Hitler." "*At the end*" could not have referred to 1942, as he said in *Inside the Third Reich*.[16]

It is worth noting also, that Speer's inclusion and account of this conversation in his prison diary may have been calculated to further ingratiate himself with his jailers, if they were reading his diary.

As Speer noted in *Infiltration*, Himmler's SS was hot on his trail due to what SS-Gruppenführer Otto Ohlendorf charged in a January 25, 1945 letter was the armaments minister's "neglect" of nuclear research and production, forcing him to make a desperate, last-minute defense that worked: enlisting his own subordinate Dr. Gerlach to verify what he'd done over the years.

The most recent statement on the Speer decision came from Adam Tooze:

> With hindsight, it is clear that the decision made by Speer and his colleagues was essentially correct. Even working with virtually limitless resources, the Americans did not manage to complete a viable atomic weapon in time for it to be used against Germany.

> But the eagerness with which the Western Allies seized on the atomic bomb at precisely the same moment that it was de-prioritized in Germany is yet more evidence of the gulf that separated the industrial and technical resources of the two sides.

> Informed by his chief scientific advisor in September 1941 that the atomic bomb program had a chance of success somewhere between one in 10 and one in two, Churchill did not hesitate to instruct the British scientists to accelerate the program to top speed... In line with the Victory Program, the Americans took the decision to accelerate the Manhattan Project even before the Japanese struck at Pearl Harbor.

> At the very least, the possibility that Germany might be working on a similar device required insurance, a kind of strategic calculation, which the Third Reich was never able to afford.[17]

My conclusion is simply this: Albert Speer—and not Adolf Hitler—dropped the ball on atomic weaponry in Nazi Germany. Hitler and Göring would have decided differently in my view.

Previously unpublished German wartime combat artwork of the versatile Ju 52/3m "Tante Ju" in flight. Used as a bomber and transport plane during the war, it entered service in 1932 as a civilian Lufthansa airliner, and is still employed worldwide, even today. Dr Speer used one for his own wartime flights. (US Army Combat Art Collection)

REICH MINISTER OF ARMAMENTS AND WAR PRODUCTION, 1942–45

"My strength always lay in organizing things ... a strength that raised me far above the limits of my potentialities."

Dr. Albert Speer, after World War II

Reich Marshal Hermann Goring (left) and architect Albert Speer on June 23, 1941 at Treptow outside Berlin inspecting exterior facade models of the proposed Reich Marshal's Office building for Berlin. The day before, Nazi Germany had invaded the Soviet Union. The two men–allies here–became rivals during 1942-45, and then enemies during the IMT trial at Nurnberg during 1945-46. (Previously unpublished photo HGA.)

Dr. Goebbels wrote of Speer in his diary:

He is the only man capable of replacing—and, one hardly likes to say so—even outdoing, the great Todt... Speer is not only approaching his huge task with the finest sense of idealism, but immense expert knowledge. He is beginning to rationalize the apparatus; will get rid of all the far too many dilettantes ... streamline the whole operation... What luck to have Speer; we work splendidly together—at last a kindred mind... Speer was an advantageous swap for Todt... Of all the leading men we have in Germany, he is one of the few whose analysis of the situation is hard, bold, and totally realistic... He is an organizational genius![1]

According to Edward R. Zilbert, the late Dr. Todt's major problem was that there was a lack of coordination among all the various Wehrmacht armaments procurements offices and agencies. He divided his time two-thirds between his construction duties and as Armaments Minister in Berlin, and the final third as administrator of the Organization Todt, which included the supervision of the building of the Atlantic Wall, and bridges in the occupied USSR.

Aircraft production had actually been reduced by 40% during the Battle of Britain, Zilbert pointed out, and there had been no new plants built by Nazi Germany throughout all of 1940 because peace was so near! This situation had continued for an entire year from June 1940 to June 1941, on the very eve of Hitler's invasion of the Soviet Union.

This then was the situation that Professor Speer faced as he took office on February 9, 1942.

Professor Speer accepted the post of Reich Minister of Armament and War Production with the agreed-upon proviso from Hitler, he asserted, that he would revert to being the Führer's architect after the war was won. That same evening, Speer took the night train to Berlin. The next day, February 9, he arrived in the Reich capital and was driven immediately to his new offices at Pariser Platz 3. "Immediately after my arrival, I paid a visit to all the important department heads in

The Great Hall of Göring's Air Ministry Building in Berlin, still standing today, but without the large Nazi eagle seen at rear. It was in this building that Dr. Speer addressed his first audience as Armaments Minister in February 1942. (Previously unpublished photograph, HGA)

A camouflaged Me 109 in the Western Desert of North Africa in 1941. The single-seat fighter carried an armament of two or three 20mm cannons and twin 7.92mm machine guns. (*Signal* magazine, Peterson)

their offices, thus sparing them the necessity of coming to me to report."[2]

Speer also addressed Dr. Todt's staff in the forecourt of the Ministry building on Berlin's Pariser Platz. According to playwright David Edgar, this is the actual text of Dr. Speer's speech that day: "Party comrades! Esteemed employees of the Ministry of Armaments! It is my sad duty to report that at the zenith of his labors, your leader Reich Minister Professor Todt was taken from you yesterday in a plane crash in East Prussia. The Führer has placed me in charge of all Dr. Todt's roles and functions. I have proposed—and the Führer has enthusiastically approved—the severest penalties for the use of materials, machinery, or manpower for unauthorized or private purposes. With the Führer's keen endorsement, I have ordered the full mobilization of up to 20 million workers from the conquered territories. I have nothing else to say. We have a war to win—and we shall win it! *Sieg Heil!*" Later, he stated: "At this moment particularly, we must not mourn and grieve for the past. Only through dogged hard work will we be able to bear our fate. The Führer's demands on the armaments industry are high, and the quotas

A German Army reconnaissance unit in Russia, 1943. (CER)

he has set are extremely hard to meet. Our output has risen from month to month, so that—unexpectedly—not only have the Führer's demands been met, but they have been surpassed."[3]

His personal secretary, Annemarie Kempf, was interviewed by Gitta Sereny for the latter's biography of Speer, and she observed some of the fundamental differences Speer brought to the Ministry. When he became Minister, as well as Kempf, Speer retained Todt's secretary, Edith Maguira, and his top aide, Karl Cliever, the trio being set up in an outer office prior to reaching his own. Annemarie Kempf said:

Speer and Todt were very different people. Todt had been in the habit of having his aides and assistants brief him on all situations. Speer never did this; he always got his information directly. He was a delegator *par excellence*, but he always remained in immediate contact with all those to whom he delegated. This was very different from virtually every other official office. All ministers were protected by assistants and aides who decided who their chief would see, and who not. Speer wouldn't have this. He opened the doors; there was complete access to him... For all those who had worked with Todt for years, this approach was very new.

Wearing a Nazi Party leader's brown uniform jacket and tie and the collar tabs of a *Reichsinspektor*/State Inspector), Speer (center) is welcomed by Reich Marshal Hermann Göring (right) at the latter's 49th birthday celebration in Berlin on January 12, 1942, as an Organization Todt general looks on at left. Less than a month later, Göring was Speer's brief rival for Dr. Todt's job. Speer stated later: "The success of our work is crucial to Germany's victory. I have vowed to the Führer to devote my strength to this end." According to author Bradley F. Smith in his 1977 work *Reaching Judgment at Nuremberg*, by June 1943, Dr. Speer got naval construction under his Armaments Ministry, and—finally—the Luftwaffe added to his production empire a year later, in June 1944. Thus, his power expanded only by stages, starting with his accession to office in February 1942. (Previously unpublished photograph, Hermann Göring Albums, LC, Washington, DC)

For those of us who had worked with Speer all along, it had become quite usual, for instance, for him to take his team along when he went to show Hitler his architectural plans. But the Ministry people were stunned, simply stunned, when he told them he wanted them along on his armament conferences with Hitler. *He* didn't care. He accepted whoever Speer wanted to bring.[4]

Hitler hated relying on experts at staff conferences and other meetings, but Speer took the exact opposite tack, routinely bringing to FHQ situation conferences a bevy of technical experts to confront Hitler's famous statistical ranting, a tactic that he termed "Hitler's tricks," also noted in his memoirs by Field Marshal Erich von Manstein.

These civilian intrusions at the military headquarters came to be termed "Speer's invasions." "Nothing could intimidate these specialists," Speer proudly asserted, and so he brought them as needed.[5]

On February 11, Speer headed the official party to receive Todt's coffin; he wrote, "The ceremony was hard on my nerves, as was the funeral on the next day in my Mosaic Hall in the Chancellery."[6] On the day of the funeral, he noted:

I was summoned to Göring. This was my first visit to him in my new capacity as Minister. Cordially, he spoke of the harmony between us

HERMANN RÖCHLING: A TYPICAL GERMAN HEAVY INDUSTRIALIST

On August 10, 1946, Albert Speer wrote a letter to Rudolf Wolters asking him to get together a collection of Speer's work, and "to set down—for the future—some of what you know of my life. I think that one day it will be appreciated." Among the possible contributors to this tome, Speer listed Hermann Röchling. Röchling was born in Saarbrücken in 1872. His life spanned four entire epochs of German history: the Wilhelminian Era, the Weimar Republic, the Third Reich, and finally, the second German Republic. In 1898, Röchling took over his father's foundry in the Saar region between Imperial Germany and Republican France.

After World War I, Röchling was sentenced to 10 years' imprisonment for spoliation of French property. He escaped the penalty by fleeing to unoccupied territory, and after a higher court repealed the sentence, he returned. Röchling managed to keep his firm free from French involvement in the French occupation of the Saar after 1918, but he lost his participation in French firms.

Röchling was an old-school German nationalist. He knew that steel would be needed in the future, given the growth of the German population and expansion of metalworking. Following his experiences in World War I, he was determined that the steel industry should do everything to ensure that Germany would have all the resources to be able to face a future confrontation. He was preoccupied by expansion of German control over Lorraine and adjoining areas, to give Germany the resources it needed, and he worked toward German reunification of the Saar territory. He met Hitler in 1933, and at Hitler's suggestion, promoted the formation of the German Front. He was a staunch supporter of the Nazi Party. After the annexation of the Saarland in 1935, he was the leading figure in the German coal and steel industry.

When World War II broke out, the Röchling family again took control of the mines that had been given up to the French. In 1940, Hermann Röchling was made Plenipotentiary for operation of the steel works in occupied Lorraine.

After Albert Speer became Minister of Armaments in 1942, the Reich Economics Ministry began urging the formation of a strong government committee to oversee private German industry steel production with one of its own men to head it. Dr. Speer noted that, "Hitler considers my suggestion of Albert Vögler acceptable, but believes that Röchling would do the job even better. Leaves the decision to us." Under Speer's ministry, Röchling was entrusted with leadership of the main ring for iron production. He was also appointed a member of Speer's Armament Council, and given the title of *Wehrwirtschaftsführer* (Wartime Management Leader). For his support, Hitler bestowed many honors on Röchling. On his 70th birthday, he was honored with the Eagle Shield of the German Reich.

Hermann Röchling was a member of the Reich Iron Association along with Alfred Krupp and Walter "Panzer" Rohland. The three men were among the most powerful industrialists in the Third Reich, and their companies supplied Hitler with most of the raw materials for his armaments.

Röchling worked hard to encourage fellow industrialists to rise to the challenge of supplying wartime Germany with what it needed, warning them that if they did not, the industry might be nationalized, an outcome he was fiercely opposed to. Röchling controlled and directed the German heavy industries in the interests of the German war effort.

In March 1945, the Saar region was taken by American troops as the Allies advanced. Röchling was confident that Germany would regain the land eventually, as they had before. By 1945, he and Speer were in agreement that to continue the war was "senseless fanaticism."

After the war, in 1947, Röchling was captured and imprisoned in Nuremberg. He was handed over to the French occupying power of the Saar region. A French court sentenced Röchling to 10 years' imprisonment, loss of civil rights, and loss of property for war crimes. His pardon and release on August 18, 1951 came about on the intervention of his former co-workers. Röchling did not live to see the restoration of his companies following the reincorporation of the Saarland into the Federal Republic of Germany in 1957.

A wartime photo of three top Nazis in conference, from left to right: German Foreign Minister Joachim von Ribbentrop, Economics Minister Dr. Walther Funk, and Reich Marshal Hermann Göring in Luftwaffe summer uniform, carrying his swagger stick baton, missing since 1945, and most likely in a private collection. (*Signal* magazine, George Peterson, National Capital Historic Sales, Springfield, VA)

while I was his architect. He hoped this would not change, he said.

On Friday, February 13, 1942, Field Marshal Milch invited Speer to a conference at the Air Ministry to which all 30 of the arms industry big business figures had already been invited. There took place a Göring–Funk–Milch conspiracy to supplant Speer, he asserted in his memoirs, but Hitler had foreseen this, and so Speer adjourned the meeting forthwith and then reconvened it that same day in the New Reich Chancellery's Cabinet Room that he had himself built. Hitler spoke to the meeting, backing up Speer. "'Behave toward him like gentlemen!'… Hitler had never introduced a minister in this way… I could do as I pleased."[7]

The Luftwaffe cabal against him dissolved instantly, or so Speer wrote. "That evening, I had a full discussion with Milch, who pledged an end to that rivalry the Air Force had hitherto practiced toward the Army and Navy in matters of procurement."[8] The Minister of Armaments and the Field Marshal became allies instead of rivals.

In the New Reich Chancellery's salon after the Cabinet Room session, Speer also got Hitler to agree to keep Martin Bormann and the Nazi Party out of his hair in personnel appointments, to which Hitler concurred on the spot, a situation that lasted until after the July 20, 1944 Army Bomb Plot explosion.

His first ministerial meeting was held on February 18 in the former conference room of the Academy of Arts in Berlin: "I spoke up for the first time," for about an hour, and his orders were backed up the next day by Hitler at FHQ Wolf's Lair.[9] To placate Göring, "I informed the German press of my appointment. To make my point more vividly, I had dug up an old photograph showing Göring, delighted with my design for his Reich Marshal's Building, clapping his hands on my shoulders."[10] That aside, on March 21, 1942—in a decree signed by Hitler—Minister Speer virtually took the entire economy away from Göring's Four Year Plan Office, in a silent Nazi coup.

Upon taking office, Minister Speer formed an immediate public relations alliance with his former boss, Dr. Goebbels, to promote

Am gleitenden Strom,
der mit dem Hochofen kommt, stubt der verantwort-
liche Minister für die deutsche Rüstung, Reichsminister
Albert Speer. Er fasst ihn in tausend Werke und
Millionen Hände, die eifriger als je die Waffen schmieden

both himself and the arms industry to the German public, primarily
in the weekly newsreels shown in theaters. "The public recognized
me as one of Hitler's most important colleagues in the war effort."[11]
To aid this effort, Minister Speer launched his own publication—
News—on March 31, 1942, that was distributed to all industry
employees.

Speer enjoyed 18 months of unalloyed power and favor as Hitler's
pet in the post of Minister of Armaments and Munitions, then War
Production, from February 1942 through November 1943, as R. J.
Overy explains:

> When Speer arrived at his post, the new rationality drive was in full
> swing... The so-called "production miracle" of 1942–44 was based
> largely on the revolution in levels of industrial efficiency initiated
> during 1941. The watchwords of the economy were announced by
> Walther Funk in April 1942: "Rationalization and Concentration."
>
> Rationalization took two forms. The first involved a more general
> rationalization and centralization of the administration and tighter
> central control over important physical resources—raw materials,
> factory equipment, and labor. The second comprised the

Left: Dr. Speer (right) talks with a Luftwaffe officer at left, as they appeared on the September 1, 1943 cover of *Die Wehrmacht* (*The Armed Forces*), the official magazine published by the OKW, in the fourth year of the war. *The Second World War in Color* noted of Professor Speer: "He now succeeded in really putting the country's economy on a war footing and in giving German industry a considerable boost by mobilizing the foreign labor force (prisoners of war and deportees), and by the systematic exploitation of the occupied countries. His action was so effective that—despite the Allied bombing and shrinking of the territory of the Reich—the production of German war industries reached its record levels in 1944."

Far left: Minister Speer (far left) uses an oversized slide to protect his eyes while he looks at molten steel being poured. Roughly translated, the caption at lower left reads, "The flowing river that comes out of the blast furnace guides the responsible one in thousands of deeds and millions of hands that are busier than ever forging weapons." (George Peterson, National Capital Historic Sales, Springfield, VA)

Right: A previously unpublished photo showing Reich Marshal Göring (right) preceding Dr. Speer (center right) after a ceremony for arms industry workers at the New German Reich Chancellery in Berlin on May Day, May 1, 1942. They are followed by Labor Front leader Dr. Robert Ley (with walking stick) and, behind him, Luftwaffe Col. Gen. Karl-Heinrich Bodenschatz. Immediately behind Dr. Ley is OKW chief Army Field Marshal Wilhelm Keitel. A proposed 1943 alliance between Speer, Göring, and Dr. Goebbels to replace the then-current trio around Hitler—Bormann, Keitel, and Dr. Lammers—fizzled out when Hitler conclusively blamed Göring for the loss of the air war. "Our retaliatory weapons have demonstrated to the whole world that German technology is still one step ahead," stated Speer. (HGA, LC, Washington, DC)

rationalization of factory practices and labor processes, and the imposition of common production standards across German industry.

An effective rationalized administration was the essential precondition for the success of reforms in distributing resources and production practices. The framework for the administration of war production was provided by the Speer Ministry.[12]

Speer recalls his plan:

I prepared a plan of organization whose vertical lines represented individual items, such as tanks, planes, or submarines... The armaments for the three branches of the service were included. These vertical columns were enclosed in numerous rings, each of which was to stand for a group of components needed for all guns, tanks, planes, and other armaments. Within these rings I considered, for example, the production of forgings or ball-bearings, or electrical equipment as a whole. Accustomed as an architect to three-dimensional thinking, I drew this new organizational scheme in perspective.[13]

The organizational set-up was rather simple: main committees, special committees for a particular product model or piece, and armaments commissions within the two above groups that were all staffed with engineers and representatives of big business. Minister Speer established the "principle of self-responsibility," whereby the State assigned the quotas to be built, and these were best achieved when left up to the manufacturers themselves, removing the prior excessive

supervision by the Nazi Party—Martin Bormann's domain. Thus, it was the industrialists themselves who controlled the production process, from cost and contracts, to fixed prices.

"Our purpose was to have everyone pull together satisfactorily," Dr. Speer asserted. Thus, "Industrial self-responsibility was launched, and, with it, the astonishingly rapid rise in arms production occurred, due to this plan."[14] Thirteen direct committees were formed for various types of weapons, and linked by 13 pools. There were also 13 commissions of Army officers meeting with major industrialists. One plant produced a single product in maximum quantity. Thus, stability was achieved, as well as mass production in place of piecemeal turn-out. "I tried to take in more and more areas of the economy."[15]

Adam Tooze explains: "The man Speer appointed to oversee the entire structure was, like Saur, a political animal: Walter Schieber [1896–1960] … a successful plant manager for I. G. Farben and an activist in the Thuringian Nazi Party, with close links to Gauleiter Sauckel…"[16] Schieber's ally was Dr. Kehrl at the Funk Economic Ministry—and another was Reichsführer-SS Himmler.

According to Overy, it was the late Dr. Todt—and *not* Minister Speer—who established the later-fabled main committees, on December 20, 1941: "Each one responsible for directing the production of a particular class of weapons or equipment."[17] These were the "rings" for which Speer later took credit. "The first Committees were set up by Todt, and then by Milch."[18] Tooze says: "Speer added little to the basic structure of industrial organizations, put in place by Todt and Milch."[19] However Overy does say that: "It was Minister Speer who added committees for ships, vehicles, airframes, aero-engines, air ammunition, and radio apparatus."[20] It was also Dr. Todt who appointed Dr. Theodor Hupfauer to survey the achieved levels of efficiency. Indeed, it was Todt who held the first main committee meeting, on February 6, 1942—two days before his fatal airplane crash. Speer noted, however, "The backing of the Führer accounted for everything." He had that, and whether Dr. Todt would have had had he not died is purely speculative.

The former post of State Secretary (the number two man in all Nazi ministries) was eliminated, and all 218 Ministry of Armaments and

A previously unpublished photo of Speer's assistant, Nuremberg Mayor Willy Liebel (left), who served as Chief of the Central Office of the Ministry of Armaments and Munitions, talking with an unknown official. "In 1944, Axis output was 40% of the number of Allied planes, but it was only 20% of Allied structure weight. Again, German and Japanese naval production by the middle of the war was concentrated on submarine output…" noted one account. (HHA)

Munitions employees and their 10,000 assistants reported directly to Minister Speer, via 30 industrial organization leaders and 10 department heads. Only Interior Minister Dr. Wilhelm Frick protested, and Hitler replaced him with Himmler in August 1943, at the very height of Dr. Speer's power.

Mayor Willy Liebel of Nuremberg headed the Planning Department, Saur was Director of the Technical Department, Dorsch ran the OT, Dr. Walter Schieber was Head of the Supply Department, and Seebauer ran the Consumer Goods Department. State bureaucrats were aghast when it was announced that department heads over 55 had to have deputies aged 40 or under.

The overall result was the doubling of production without increasing equipment or labor costs during the period February–August 1942. Opposed by the Nazi Party, Speer nonetheless "Applied the ideals of democratic economic leadership"—and succeeded.[21] Paradoxically, the United States went in exactly the opposite direction at the very same time! Mistakes were admitted, criticism was allowed, and, thus, improvements were made. Fear of factory heads of their regional Gauleiters was removed. Indeed, Sauckel wanted to *shoot* some factory heads. "We owed our success to technicians… I preferred uncomfortable associates to compliant tools," Dr. Speer noted.[22] He also assured the captains of industry that his wartime measures were only temporary, until war's end.

"At the end of 1942, the telephone book of the Speer Ministry listed 249 committees, subcommittees, rings, and sub-rings."[23] And thus Dr. Speer *did* create his own Nazi bureaucracy, contrary to what he alleged in his trio of postwar tomes.

In 1943, Speer established his armament commissions, to bring greater coordination between weapons design and production needs. In the spring of 1944, the Ministry of Armaments and Munitions took over aircraft production from the Luftwaffe at Field Marshal Milch's urging and instigation, and immediately reduced the number of separate models from 42 to five. Earlier, in 1942, he reduced the

A presentation ceremony at the Platterhof Hotel on the Obersalzberg in Bavaria on February 6, 1943 during which the grateful Führer gave Dr. Speer the first Fritz Todt Prize, or—as it was formally known—the Fritz Todt Ring of Technical Achievement. In the background stands Walter "Panzer" Rohland and other top members of Speer's working staff. According to Domarus, "The Führer expressed his heartfelt gratitude and total appreciation to the participants. He said that these outstanding successes had to be attributed primarily to German technology's being invigorated by the leadership of Reich Minister Albert Speer and by new ideas, as well as to industrial responsibility being energetically controlled by the Führer himself." The ring came in an impressive casket fashioned in amber and silver. On its lid was a silver plaque bust of Fritz Todt. Around the sides were emblems of technology. On the oval face was a large 'T'. This face was surrounded by oak leaves. The body of the ring was interlined with interlinked swastikas. Added Domarus: "This peculiar award was never heard of again. The pin now issued consisted of the DAF gear wheel with the swastika and an eagle holding a banner with the inscription, 'Dr. Fritz Todt' between the outstretched claws." (HHA)

number of truck models made from 151 to 23, as he also did for special-purpose machine tools, and the number of the workers who used them. All of this resulted in a 100% rise in production during 1941–44, according to Overy.[24]

Wages were increased, with special "Speer recognition" bonuses for achievement given as well. Foreign worker output was but 50–80% of that of German workers, however, while, "By 1943, the bulk of the industrial workforce was concentrated in arms production... The production of all weapons was expanded by 130%, aircraft by over 200%, and tanks by over 250%."[25] These figures were the key to the "Speer miracle," indeed, its very heart and soul.

Ever the innovator willing to push the Nazi production envelope as far as it would go, in 1943, Dr. Speer appointed a car builder to manufacture U-boats faster, and a single submarine's production was reduced from 42 weeks to 16, and accomplished by breaking down production into eight separate sections, each produced in bulk at different plants and then assembled at a central yard, using "layout and flow of production," as practiced by Ford of Germany at Cologne.

Also introduced was automatic machine-pressing and stamping of machine parts, previously done by hand. "Savings engineers" were employed for better raw materials usage. "Weapons were designed to last for 25 years." In fact, their quality was considered to be *too* high. There was, too, "Better use of labor on the shop floor... Saur singled out special tools as the most important single factor in explaining efficiency gains."[26]

A weapons demonstration of a 7.5cm PAK antitank gun in the field with, from left to right, Army Chief of the General Staff Col. Gen. Kurt Zeitzler, Speer, Field Marshal Keitel, Hitler, and Dr. Saur, Chief of the Technical Office, and, as such, deputy to both Drs. Todt and Speer. Hitler thought very highly of Saur, and at the end of the war—in his Last Will and Testament—replaced Speer with his former deputy. Born on June 16, 1902 in Düsseldorf, Saur was the third child in an upper-middle-class Lutheran family. He received a degree in iron and steel engineering, and was doing well in that field when his father suddenly died in 1937, and he was forced to take over the family firm of 400 employees at age 25. When it went bankrupt in 1929 during the Great Depression, Saur's family never forgave him—even after the end of World War II. He went to work for industrialist Fritz Thyssen in the Ruhr, an early financier of the Nazis, and ran the office when the boss was away. When he discovered Thyssen's huge pornography collection, the staid Saur was shocked. He joined the Nazis on October 1, 1931, against the wishes of his family. He married his wife Veronica, 12 years younger than him, in 1936, and went to work for Dr. Fritz Todt that same year. The two hit it off from the start. The Saur family numbered five children by 1944 (a sixth child, a girl. died eight weeks after birth). Like the Bormanns, the Saurs were both non-religious and authoritarian. In his Spandau diary on September 26, 1948, Speer wrote: "Saur was the strongest advocate of coercive measures for increasing production, a ruthless slave driver on the industrial front. In the final phase, his blatant opportunism caused me serious difficulties. Now he is serving the prosecution; he has remained the same man. I would not wish it for anyone, but if one of my close associates has ever deserved to be tried, Saur is the man." (Previously unpublished photo, HHA)

Using effectively his newly gained rapport with Grand Adm. Karl Dönitz, Minister Speer got all naval production within his domain in 1943, but "There was a shortage of engineers who understood the new methods."[27] Ironically, German workers were more resistant to rationalization than were their foreign counterparts. Not surprisingly, though, it was Martin Bormann's Nazi Party Gauleiters who proved to be Minister Speer's most persistent opponents within the state, and not Himmler's SS (who admired his efforts so much that they wanted to absorb his organization into their own—entirely).

As Professor Overy noted, Minister Speer used existing plants rather than building new ones. After having visited 20 plants in the Berlin area in a single night and finding none operating, he established double shifts the very next day. In another instance, he increased a mere 180 Me 109 fighter planes being made at seven different locations to 1,000. Aluminum stock wastage was also corrected in the early days of the new Speer regime. He relied on engineers because they were "Trained in scientific management techniques."[28] Most building construction of all kinds was suspended in 1942, especially his own, much to Hitler's chagrin.

American-style automation was introduced, whereby the product being constructed was brought to the stationary workers, rather than having teams of workers moving from site to site, the prior practice. Finishing time was also reduced, resulting in "longer production runs and fewer interruptions" of assembly lines, as previously.

To achieve so much in his first six months as Minister, Speer committed himself completely to his task. His secretary Kempf said "Until he collapsed in January 1944, Speer worked like a demon. His day always started before dawn, never ended until deep into the night, which—from the very start of his ministerial appointment—he

SPEER: ASSASSIN?

At Nuremberg, Dr. Speer asserted that as the war turned sour, he had tried to assassinate Hitler and the other occupants of the Führer Bunker underneath the NRC in Berlin with poison gas during the waning days of the Third Reich, early in 1945.

It is also interesting to note one of Hitler's first reactions after the German Army-inspired bomb blast at Rastenburg on July 20, 1944, as stated by Dr. Schmidt: "Mention was made of Hitler's suspicion that the would-be assassin was a Todt Organization employee working in Rastenburg."

As head of the OT, Dr. Speer himself was responsible for all the employees working on the fortification of the bunker installations at FHQ. Indeed, the very building in which the explosion had occurred was in Dr. Speer's own briefing hut—the so-called Speer Barrack—a fact not overlooked by Dr. Goebbels in the aftermath!

The Minister of Armaments therefore felt implicated, especially since he could not tell Dr. Goebbels what kind of security screening the workers had undergone when they were being selected for work within the confines of FHQ.

According to Dr. Schmidt,

> Throughout the day, Goebbels was very distrustful of his fellow minister. Consequently, he was lying when he claimed, in his speech of July 26, 1944, that he had instantly known that none of his

A dangerous figure for Dr. Speer during 1943–45 was SS Gen. Ernst Kaltenbrunner (1903–46), seen here second from left with Reichsführer-SS Himmler (second from right). A deputy to the Reichsführer-SS, Dr. Kaltenbrunner also reported directly to Hitler, a distinction previously achieved by the assassinated Heydrich. Like Göring and Dr. Goebbels, Kaltenbrunner, too, had ambitions of some day succeeding Joachim von Ribbentrop as German Foreign Minister. Recalling General Kaltenbrunner's unexpected visit after July 20, 1944, Speer

A previously unpublished photo taken at Rastenburg following the unsuccessful Army bomb plot to kill Hitler in July 1944. From left to right: Reich Marshal Göring shakes hands with Grand Admiral Dr. Raeder, while the new Chief of the Army General Staff—Col. Gen. Heinz Guderian—looks on as an SS man gives a Hitler salute. (HHA)

recollected: "Without preface he began: 'In the safe at the Bendlerstrasse, we have found papers relating to the government the July 20th men intended to set up. You are down on the list as Armaments Minister.'... In pencil, a skeptic had written alongside, 'If possible'—and added a question mark. That unknown officer—and the fact that I had not accepted the invitation to the Bendlerstrasse on July 20th, saved my life." (Previously unpublished photo from the Heinrich Hoffmann Albums, US National Archives, College Park, MD)

construction workers at Führer Headquarters could have committed the crime.

Meanwhile, Hitler had gotten information about the conspirators, and he gave up his suspicion about the OT workmen on the very same day. Mussolini visited him in Rastenburg as scheduled on July 20, and when Hitler was taking him back to the railroad platform, he stopped to talk to the construction workers, who had been arrested.

He told them that they were free now: "I knew from the very first that *you* weren't the culprits..."[29]

Still, the question of Goebbels' suspicions of Dr. Speer persists—did he have a *reason* for it? Was Professor Speer implicated in the July 20 plot?

A prewar German inland waterway canal. Dr. Speer inherited authority over these from the late Minister Dr. Todt in 1942. (Previously unpublished photo, HHA)

very often spent in his office when he was in Berlin, sleeping perhaps three hours on a cot. He very rarely saw his family. The Speers had a small townhouse in a residential suburb, Berlin-Schlachtensee, which he had built in 1935, but when the family stayed there, he basically never saw them, especially not the children..." His wife concurred: "The children hardly knew him... For all intents and purposes, the children did not have a father."[30] He recalled later:

> Of course, my job as Minister of Armaments used all my reserves of strength. I had to throw myself into the job with an excess of personal involvement in order to make up for my lack of overview, my inadequate knowledge. From morning to night, even during hasty meals, I held important talks, dictated, conferred, made decisions. From conference to conference the subjects jumped around from one problem to the next; I frequently had to find temporary solutions, make decisions of vast import. I suppose I stood the pace only because I would take a trip every two weeks, spending a few days visiting bomb-damaged factories, headquarters on the military fronts, or building sites in order to gather fresh impressions, to come into contact with practical activities. These tours represented more work, of course, but they also gave me fresh energy. I loved to spend myself to the very limits of my strength. In this, I was basically different from Hitler, who regarded the constant activity imposed on him by the war as a terrible burden; he was forever longing to return to the easy going pace of earlier years.[31]

In his 1971 *Playboy* interview with Speer, Eric Norden stated: "By all accounts, your successes—at least in the early days—were quite phenomenal." Dr. Speer answered:

> Yes, once we had restructured the war economy under a Central Planning control and mobilized our industrial reserves and resources, the production results were remarkable. Within six months of my appointment, production had soared in every area under our control.
>
> In the period from February to August 1942, production increased by a ratio of 27% for guns and 25% for tanks, while ammunition output rose by 97%. Our armaments production in that half-year rose by 59.6%. By 1943, our factories were producing seven times the weapons produced in 1942, and over five times the amount of ammunition; our total munitions production rose from 540,000 tons in 1941 to 2,558,000 tons in 1943.
>
> Even as the Allied air onslaught grew in ferocity, our production figures continued to skyrocket. The Allies were quite amazed by the way our arms industry kept going almost to the end... The forces arrayed against us were too overwhelming for even the most brilliant industrial planning to overcome.[32]

R. J. Overy's survey of the wartime economy of Germany presents a slightly different perspective on the figures: "The largest part of Germany's industrial workforce for war work and industrial capacity was mobilized for war work in the first two years of war; there was an increase of 149% between 1939–41, but an increase of only 11% between 1941–43."[33]

Richard Grunberger summarizes Speer's achievements as Minister:

> Speer's cadre of 6,000 honorary administrators—drawn from the entrepreneurial and managerial classes—was responsible for a staggering increase in arms output. Its creator has subsequently and convincingly stated that, but for this productive spurt, Germany could not have continued the war after 1943...
>
> By 1944, the average working week was 60 hours, and 72 hours in priority arms industries. On the whole, labor discipline remained high, so high in fact that the output of armaments increased 230% between 1941–44 (during which time the workforce involved expanded a mere 28%). Albert Speer—who masterminded this wartime economic miracle—has reason to claim that he helped to stave off Nazi Germany's defeat by two years...[34]

Luftwaffe paratroopers aboard a Kettenkrad with trailer in Tunisia in 1943. (US Army Combat Art Collection)

Norden asked Speer about his achievements: "…you did succeed in delaying the inevitable; without your efforts, according to some historians, Hitler would have had to admit defeat as early as 1942 or 1943. Hundreds of thousands of soldiers and civilians on both sides died in that period, and yet you still speak of your achievements with apparent pride." Professor Speer responded,

> I am not proud of my role in prolonging the war—just the opposite… In those years, I did not think in ethical or humanitarian terms. All I was interested in was increasing our war production. I cannot help feeling stirrings of pride… Those were the days of my youth, and I

Hitler with Dr. Herbert Backe (1896–1947), Minister of Food (1942) and Minister of Agriculture (1944) in Berlin. Born at Batum in Georgia, Dr. Backe was captured by the Allies after the war, and committed suicide by hanging in his cell at Nuremberg on April 6, 1947. (Previously unpublished photo, HHA)

Two Third Reich ministers chat outside an above-ground bunker at FHQ Wolf's Lair, Rastenburg, East Prussia during the war. At left is Professor Speer, and at right his colleague, Minister of Posts (that included being Director-General of the State Railways) Dr. Julius Dorpmüller (1869–1945). Noted von Below in his postwar memoirs: "The last few, quiet weeks of the war … Hitler was not deceived. He knew that Speer was no longer convinced of victory." Von Below had been Hitler's adjutant since 1937, and on May 22, 1944, began pulling "double duty," serving as Speer's aide also for the final year of the war. According to von Below—who served as wartime adjutant to both Hitler *and* Speer simultaneously—the latter reported to the former at least every two weeks on armaments, while Speer himself insisted that he was often in the Führer's presence daily "for years." (Previously unpublished photo, HHA)

achieved things which many people predicted were impossible, and I suppose my ego still takes pleasure in those accomplishments. Then I think of all the cities destroyed, the soldiers killed, the Jews butchered between 1943–45—and my pride turns to sickness.[35]

When asked in 1971 when he thought the war had turned against the Germans, Dr. Speer answered,

A German Army Pz.Kpfw III tank in Russia, 1941. Like Dr. Todt before him, Professor Speer, also thought that the war was lost, *before* Stalingrad. "We were very relieved to learn that the English had remained far behind our developments in tank construction. Thus, we had time to test our new Panther tank in peace, and put it into service just right!" That occurred the following year at the Battle of Kursk in July 1943. (CER)

The final and irrevocable turning point was Stalingrad, although things had begun to go badly even before that. Hitler never expected a prolonged war, and, therefore, he had never prepared for one. The moment his series of blitzkrieg victories ceased and the struggle began to drag out, we encountered shortages of strategic raw materials— particularly fuel—that seriously hampered our war machine.

It really was a miracle that we were able to keep going as long as we did, considering the huge drains on our resources and manpower. These problems were manifested particularly acutely in my own armaments production work.[36]

Tooze discusses this crucial phase of the war: 1942 was the pivotal year, as both Speer and Field Marshal Milch realized all too well, as only a military victory in Russia by Army Group South over the Red Army could have won the war economically, by giving the Reich the areas that held the raw materials its war machine needed; but that didn't happen, and thus the war was lost. "In 1942—in the first full flush of the 'arms miracle'—Germany was considerably out-produced by the extraordinary mobilization of the Soviet economy. This Soviet

effort was unsustainable. By 1944, Germany had caught up with the Soviet Union, but the summer, autumn, and winter battles of 1942–43 were the key to deciding the war on the Eastern Front," and the Reich lost most of them.[37] Thus, the war was lost militarily by the time of the Battle of Kursk in July 1943, almost a year *before* the Allied D-Day invasion of Northwest Europe at Normandy, France.

In 1943, Minister Speer wanted to introduce his German domestic front "ring" industrial system of various segments within the overall whole to all of Europe, making of it, therefore, a pan-European economic regime: "one for coal, another for steel, still others for machinery, or electrical engineering." In *Spandau*, he quotes Hess as saying—"pensively"—"You're not so dumb!"[38] Both Hitler and von Ribbentrop blocked it, being as they were German nationalists first, last, and always.

On September 2, 1943, Dr. Speer was promoted by Hitler from being Reich Minister of Armaments and Munitions to Reich Minister of Armaments and War Production, with military adjutants as a counterweight to the Nazi Party power wielded by his arch-enemy, Martin Bormann. The Army's Personnel Director and also Hitler's chief Wehrmacht adjutant, Major General Rudolf Schmundt, asserted: "You can always rely on the Army, Herr Speer. It is behind you," and opposed to Göring after the Stalingrad airlift debâcle that doomed the German 6th Army.[39]

Slyly, Hitler also played off Minister Speer against Himmler when both entered his conference room together one day, referring to them as "You two peers," leading Chief of the General Staff General Kurt

A previously unpublished photo of Martin Bormann taking a picture with the Leica camera belonging to Reich Photo Reporter Professor Heinrich Hoffmann at one of Hitler's 1940 FHQs. In August 1943, Wilhelm Frick was replaced by Himmler as Minister of the Interior, and the latter also took over command of the RAD. Speer's longtime nemesis was Reichsleiter Martin Bormann (1900–45), named Secretary of the Führer in May 1943. On September 2, 1943, Hitler issued decrees that outlined Dr. Speer's powers: "The Todt Organization will be an institution geared toward the implementation of all kinds of tasks involving construction and building, tasks of decisive impact on the war. The chief of the Todt Organization will be the Reich Minister for Armaments and Munitions. He will answer directly to me, and will be responsible only to me. The deployment of the TO will take place in the Greater German Reich and in the annexed or occupied territories. Deployment will be ordered by the chief of the TO… Implementing regulations on the structure of the TO will be issued by its chief." (HHA)

Martin Bormann's wartime Nazi Party administrative empire: the *Gaue* (Regions) that allowed Hitler, Hess, and then Bormann to effectively rule the Third Reich until the very end of the war and the regime. On May 1, 1944, Hitler named several Pioneers of Labor, according to Domarus: "The Reich Minister of Transportation, Dorpmüller; the Saarland Councilor of Commerce Röchling, the aircraft designer and manufacturer Professor Dornier, and the industrialist Albert Vögler." (LC)

Zeitzler to mention that "A new sun has arisen after Göring," and each man felt that he meant him. This exacerbated the already apparent rivalry between Speer and Himmler.

Regarding the growing RAF air raids, Speer recalled: "In spite of the losses of factories, we were producing more, not less."[40] As for civilian morale under the ever-increasing weight of Allied bombs, there was "growing toughness." By August 1943, Speer wrongly believed that war industry was "collapsing" due to the Allied bombing, but Hitler predicted that, "'You'll straighten all that out again.' In fact, Hitler was right."

On September 12, 1943—with the Italian switching of sides to the Allies—Speer gained industrial northern Italy as part of his overall empire. Five days later, French Minister of Production Jean Bichelonne made a State Visit to Berlin officially to aid Minister Speer in what both of them wanted: a Nazi European Common Market. But Sauckel protested as a German nationalist, and Hitler quashed the idea.

In late 1943, Speer began to have serious, conclusive doubts about the war's winnability. Speaking to Norden, he said:

> The Allies didn't know it, but they had victory within their grasp in 1943. In August of that year, they began a series of devastatingly

FRITZ SAUCKEL (1894–1946)

Born on October 27, 1894 at Hassfurt in Lower Franconia, Ernst Friedrich Christoph Sauckel was forced to cut his early schooling short when his mother fell sick, and joined the merchant marines of Denmark and Norway at 15. Having sailed the world, Sauckel was aboard a German ship at the outbreak of World War I on his way to Australia when he was captured, and then interned in France from August 1914 through November 1919, fully a year after the Armistice that ended the fighting.

Back in the defeated Weimar Republic, Sauckel went to work in a factory at Schweinfurt, while studying engineering at Ilmenau during 1922–23, the same year that he joined the Nazi Party as Member #1,395. In 1924, Sauckel married Elizabeth Wetzel, daughter of a German Social Democrat and the couple had 10 children.

Sauckel remained a member of the dissolved Nazis during 1924–25 after the failure of the Beer Hall Putsch/Revolt, and publicly rejoined in 1925 when Hitler reorganized it. The Führer named Sauckel as Gauleiter of Thuringia in 1927, and two years later he also became a member of its regional government.

When Hitler was named Reich Chancellor in 1933, Sauckel was appointed Reich Regent of Thuringia, and was also an elected member of the Reichstag in Berlin. In 1934, he was awarded honorary generalships in both the SA and SS in addition.

Sauckel began World War II as Reich Defense Commissioner for the District of Kassel before the Führer named him as General

Below left: German wartime slave labor tsar Gauleiter/Regional leader Fritz Sauckel seen as an accused war criminal at Nuremberg. (US Army Signal Corps photo by Ray D'Addario.)

Below: Dr. Speer (right) test drives a new tank chassis. Up and behind him sits his rival, Dr. Ferdinand Porsche the Elder, wearing cap with wind goggles on the bill. (*Signal*, LC)

Below: Fritz Sauckel (left) with his Four Year Plan boss, Hermann Göring (center, in whites) before the war. (Previously unpublished photo, HGA.)

Plenipotentiary for the Employment/Allocation of Labor on March 21, 1942, as the hand-picked candidate of Martin Bormann to thwart the selection of the man whom Dr. Speer supported: his old-time Berlin Party colleague Karl Hanke. This appointment was rightly seen within the Party as a victory for Reich Leader Bormann and as a defeat for Professor Speer.

Although nominally under Göring's Four Year Plan—as was, indeed, also Dr. Speer—Sauckel in reality was Speer's *co*-worker who reported directly to Hitler, as did Speer.

Of the required five million workers brought to Nazi Germany by Sauckel to fulfill Speer's wartime manpower quotas, only 200,000 came voluntarily, the rest as conscripted laborers. Conditions for the forced laborers were appalling, particularly in the Ruhr coalmines and Krupp armament work camps. Hours were long, pay was abysmal, and serious illnesses such as typhus wrought havoc among the undernourished workers.

According to the formal Allied indictment against Defendant Sauckel before the International Military Tribunal at Nuremberg during 1945–46:

The defendant Sauckel between 1921–45 was: a member of the Nazi Party, Gauleiter and *Reichsstatthalter* (Governor) of Thuringia, a member of the Reichstag, General Plenipotentiary for the Employment of Labor under the Four Year Plan, joint organizer with the defendant Dr. Ley of the Central Inspection for the Care of Foreign Workers, a general in the SS, and a general in the SA.

The defendant Sauckel used the foregoing positions and his personal influence in such a manner that he participated in the economic preparations for wars of aggression, wars in violation of treaties, agreements, and assurances set forth in Counts 1 and 2; he authorized, directed, and participated in the war crimes set forth in Count 4, including particularly the war crimes and crimes against humanity involved in forcing the inhabitants of occupied countries to work as slave laborers in occupied countries and in Germany.

At Nuremberg, Sauckel defended his labor law as having "Nothing to do with exploitation. It is an economic process for supplying labor." Sauckel lamely denied that it was also slave labor, or that it was commonplace to work these forced laborers to death under the alleged "extermination by labor," or to mistreat them, as the court found was done. As the convicted, "Most notorious slave-driver in history," Sauckel went to the gallows—some said in Speer's place—on October 16, 1946. His last words before his hanging were, "I'm dying innocently! My sentence isn't just. God protect Germany!"

effective bombing raids on our ball-bearing factories, which were pivotal to all our armaments production. By paralyzing this crucial nerve center, they could have completely wiped out our armaments production.

Ball bearings were vital and irreplaceable components of every major weapon from tanks to planes, and a few more Allied raids would have crippled our arms production; the war would have been effectively over. And then—to our utter astonishment and tremendous relief!—the Allied raids abruptly terminated, and we were able to rebuild our damaged factories, although they never again achieved their full industrial capacity.[41]

Allied bombing meant the gradual collapse of the overall supply system and also the transportation establishment. Despite this, however, Germany acquired 55,000 new combat aircraft in 1944, and 30,000 tanks. The pending overall collapse after January 1945 occurred primarily due to the Allied bombing campaign, when 36,000 tanks had been projected for construction in the final year of Speer's "production miracle."

The Allied raid on Berlin of November 22, 1943 destroyed the Speer Ministry Building on the Pariser Platz: "In place of my private office, I found nothing but a huge bomb crater… Our ability to produce more in spite of the air raids must have been one of the reasons that Hitler did not really take the air battle over Germany seriously." Politically, "many of us believed that Hitler would end the war at the right time."[42]

On the other hand, he later came to believe that the economic turn of the war's tide occurred not at Stalingrad or even at Kursk, but in November–December 1943. He also predicted that, "The war would be over 10 months after the loss of the Balkans."[43]

Still, he made monthly phone calls to Hitler at FHQ regarding production figures: "A word from Hitler had lost none of its magical

Walter Funk (1890–1960), seen in a previously unpublished portrait by German artist Erler. Funk was Professor Speer's colleague as Reich Economic Minister and President of the Reichsbank. (US Army Combat Art Collection, courtesy former Director Marylou Gjernes)

A previously unpublished photo of a Luftwaffe Flettner Fl 282 Kolibri (Humming Bird) experimental helicopter at Rastenburg being inspected by Hitler and his entourage during the war. From left to right are seen Chief of Staff German Army Col. Gen. Kurt Zeitzler, unknown Army officer, Reichsführer-SS Himmler, two unidentified men, Minister Dr. Speer, unknown man, Hitler, and other unknown men. Field Marshal Wilhelm Keitel stands sixth from right in the central group in left profile. At far right, the man in the civilian dark hat and overcoat is most likely Speer's man, Dr. Theodor Hupfauer. (HHA)

German Army infantry and a Sturmgeschütz III mobile assault gun advance into the USSR on September 19, 1942. "After mid-1942, arms production concentrated on tanks and airplanes. Within six months, the total number of tanks produced increased by 25% and the total number of airplanes by 60%, with the labor force only increasing by 30%. Speer said that these results came from his policy of giving technicians responsibility for entire sections of the armaments industry on the grounds of their professional skill, rather than on the grounds of Party membership." (Reich Railroad, LC)

force," he recalled wistfully. On the other hand, Hitler grew sharp with him when challenged openly: "*I am chairing this meeting and will decide, not you!*" When Minister Speer sent Zeitz a memo stating that there was enough steel in reserve to last until December 1944, Hitler angrily issued an order on November 13, 1943 at FHQ stating that there would henceforth be "No memos to anybody but me!"[44] Their old bond, strained by the pressures of the war, was beginning to crack.

From fall 1943 onward, Hitler began more and more to call Speer's deputy, Karl Otto Saur, for the monthly Ministry of Armaments production figures instead of his own appointed Reich Minister. Meanwhile, both Saur and Xaver Dorsch—covert cronies of Speer's mortal political enemy, Reichsleiter Martin Bormann—made their chief

German armored units in action in North Africa, 1942. (Previously unpublished, US Army Combat Art Collection)

feel "insecure in my own Ministry."[45]

On October 6, 1943, Speer gave a 50-minute address to the top officials of Nazi Germany assembled at a castle in Posen, Poland on the critical state of the war. Dr. Josef Goebbels noted in his diary: "Speer told them very bluntly that no protests and no arguments would deter him [from converting all plants to war production.] He is, of course, right..."[46] But that is not how the "Golden Pheasants" of the Nazi Party—the Reichsleiters and Gauleiters (District and Regional Leaders)—and Secretary to the Führer Martin Bormann saw it, as they sat stunned in the sumptuous Golden Hall of the castle. In particular they took umbrage at the words they rightly saw as a direct threat to their domains: "The manner in which some of the Gaue have hitherto obstructed the shutdown of consumer goods production will no longer be tolerated... I am prepared to apply the authority of the Reich

Above left: Grand Admiral Dönitz (right) receives the salute of one of his officers during a wartime naval review. Dr. Speer took over all U-boat construction in 1943 at the behest of Admiral Dönitz. (CER, LC)

Above: Previously unpublished shot of Hitler (right) receiving victorious U-boat commanders in audience at FHQ Wolf's Lair, Rastenburg, East Prussia, 1943. (HHA)

Interior of Speer's bombproof U-boat pens at Trondheim, Norway. (CER)

Dr. Speer's wartime bombproof concrete U-boat pens at Trondheim, Norway, where Hitler wanted to build a superbase port city by 1950. The U-boat at left is a Type VII, and that at right is a Type IX. (CER)

Government at any cost. I have discussed this with Reichsführer-SS Himmler, and from now on districts that do not carry out within two weeks the measures I request will be dealt with firmly."[47] Bormann returned to Hitler with these words in a successful attempt to undermine Speer's standing with his Führer.

Himmler also spoke at Posen. This was the notorious occasion on which he told the assembled guests about what the SS had been doing "in the East" to the Jews and others since the German invasion of the Soviet Union on June 22, 1941; this was "Part of Hitler's determination to make sure that his supporters were all implicated in the catastrophe he was bringing on Germany," according to Gitta Sereny.[48]

Speer later claimed that he wasn't there—that he left before Himmler spoke—and that therefore he didn't know about the terrible realities of "The Final Solution of the Jewish Question." He did, however, know of the slave labor conditions in use at the underground rocket factory at Dora that was under his direct control.

Joachim Fest said of Speer: "The Posen Conference marked the beginning of his descent," for two reasons: first, he had challenged the

SPEER AND THE HOLOCAUST

Dr. Speer, interviewed by Eric Norden in 1970, was asked: "When was it finally decided to annihilate the Jews?" Speer replied:

> I am sure that Hitler had it in his mind ever since *Kristallnacht* [the night of November 9/10, 1938], but I learned from evidence introduced at Nuremberg that the actual decision was made at the Wannsee Conference in [January] 1942, once he knew the war was going to be total, with either absolute victory or absolute defeat at the end.
>
> I think that knowledge eliminated the last remaining political and diplomatic restraints and liberated his most terrible instincts. It wasn't a ministerial decision: most of Hitler's government associates— including me—never even knew about it till they were told at the end of the war... I know that many people outside Germany believe that everyone in the country knew of the extermination, but that just wasn't the case, as historians of the period will tell you.

Earlier, Norden asserted that it was "An extermination you did nothing to prevent and—by successfully prolonging the war as Armaments Minister—actually assisted." Dr. Speer replied:

> ...It was my duty to *confront* it, to assert my individual and collective responsibility for it. That was my greatest failure. From the very beginning, I should have seen where Hitler's hatred of the Jews would lead... I accommodated myself to his mania ... that he would abandon when he came to power... [Later] I just stood aside and said to myself that as long as I did not personally participate, it had nothing to do with me... My toleration of the anti-Semitic campaign made me responsible for it... On the day after Kristallnacht in 1938 ... I was bothered only by the litter... Hitler crossed a Rubicon; as barbarous as his treatment of the Jews had been, I don't think even he had contemplated their physical extermination until then. More was shattered than glass that night...

The photo that might have gotten Dr. Speer hanged, had it been available to the IMT in 1945-46. According to John K. Lattimer, MD, ScD, "It is the only record of his ever visiting a concentration camp (Mauthausen), where he was invited to accompany Nazi Gauleiter August Eigruber on a 'VIP' tour, where they were shown nothing bad. Eigruber (center, left) was hanged after the war, and Speer is seen at center right. Other writers feel that Speer deputy Fritz Sauckel was hanged in his rightful place. (LC)

Wearing the same style striped prison uniforms, slave laborers work on a V-2 rocket engine at Dora in the Harz Mountains during the war. (Walter Frentz.)

In the summer of 1944, I was visited by one of my old friends, Karl Hanke, the Gauleiter or District Governor, of Lower Silesia... On this occasion, he came into my office and just slumped down into my green leather armchair, and was silent for a long time... He told me he had just visited a concentration camp in Upper Silesia and he urged me in a faltering voice never—under any circumstances whatsoever—to accept an invitation to inspect that camp. He had seen horrible things there, things he was not allowed to discuss—things he could not bring himself to discuss. I had never seen Hanke in such a state ... I did not investigate. I did nothing. Hanke, of course, was speaking of Auschwitz. From that point on, I had irrevocably condemned myself. My moral contamination was complete. I still feel directly responsible for Auschwitz in a completely personal sense.

Norden asked him: "But what would you have done if you *had* known that 6,000,000 Jewish men, women, and children were being exterminated?" Speer answered: "Would I have fought, protested, tried to stop the slaughter, risked my life? I must say I doubt it. Looking back over the decades at the man I was then, I can expect no moral courage from him."

Norden countered: "If Hitler had admitted to you that he was annihilating the Jews, what would you have said to him?" Dr. Speer replied, "... I would have said, 'You are killing them? That is insane! I need them to work in our factories.' That would have been, I am afraid, my first reaction at the time." Norden rejoined: "No moral outrage? No revulsion?" and Speer answered: "The man who left Spandau was not the same man who entered... I had no thought other than oiling the war machine, even with blood."[49]

In his 1997 work *The Good Nazi: The Life and Lies of Albert Speer*, Dan van der Vat stated that, "It is simply impossible to believe that Speer was unaware of the Final Solution after Himmler's Posen speech in October 1943, if not considerably earlier." Not only that, but Speer and Dr. Rudolf Wolters—his chief deputy—conspired both during the former's detention in Spandau, and after he was released, to keep hidden from prying, independent eyes the wartime office chronicle. This was to ensure that no one found out about Speer's eviction of 75,000 Berliners from their "Jew flats" to make way for the Führer's grandiose planned Germania. The fate of these people was clear: they were shipped "to the east," i.e. Auschwitz. In addition, van der Vat concluded—and I concur—that Dr. Speer also knew all about the thousands of deaths of slave laborers imported into the Reich's secret underground V-1 and V-2 factory, Dora, and began as early as April 1944 to sanitize his postwar reputation to survive into the non-Nazi era: "The war was lost, and a terrible day of reckoning was approaching."[50]

A captured Kriegsmarine (German Navy) officer (center) of the U-585 in captivity. By the late summer of 1943, the U-boat menace was being brought under Allied control at sea. (US Navy photo)

Gauleiters' power, and, second, he had realized the dire peril that he was in regarding "the Jewish Question" should the Third Reich lose the war—as he and Hitler both now knew it would.[51]

In the midst of these two unsettling dilemmas—and his almost by now certain knowledge that Germany had lost the war in terms of production against the Allies—Speer decided to spend Christmas 1943 in German-occupied Lapland with his personal secretary, Annemarie Kempf and one of his top aides, Rudolf Wolters—rather than with his family or Hitler.

It was there that he developed a swollen left knee, which left him in a state of collapse and overwork: "Physically, I was nearly worn out at the age of 38. I was taken to a hospital on January 18, 1944."[52]

He asked his friend SS Dr. Karl Brandt—Hitler's own surgeon and Commissioner for Public Health—for advice, and the latter recommended to him SS Dr. Karl Gebhardt, a leading orthopedic surgeon with a hospital of his own outside Berlin, and a personal friend of Himmler's.

Speer later claimed that he did not know that the hospital at Hohenlychen was an SS facility, but I find this difficult to believe in light of his detailed knowledge of virtually everything in Nazi Germany, and also due to the fact that—after the war—it was revealed that SS criminal medical experiments were performed there.

Speer was now in the clutches of the SS, and another of his rivals, Reichsführer-SS Himmler, who was certainly plotting during 1944, planning to inaugurate an SS State with himself as Führer in the spring of 1945 in an alliance with the Western Allies to continue the war against the Russians. For his plans to succeed, Speer had to go. If the new patient at Hohenlychen were to die conveniently under SS medical care, Himmler could then concentrate next on the man closest to Hitler: Martin Bormann.

On May 10, 1941, Bormann had succeeded Deputy Führer Rudolf

Right: The German High Command in March 1943 for the annual Heroes' Memorial Day commemoration outside the War Memorial (Guard House) in Berlin. From left to right: Reich Marshal Göring, Chancellor Hitler, OKW chief Field Marshal Keitel, Navy C-in-C Grand Admiral Karl Dönitz, Reichsführer-SS Himmler (representing the Waffen/Armed SS for the first time), and Luftwaffe Field Marshal Erhard Milch. Both Drs. Todt and Speer dealt with all these men at close quarters throughout the war. (Previously unpublished photo by Prof. Heinrich Hoffmann, courtesy of the US Army Military History Institute, Carlisle Barracks, PA.)

Left: In this previously unpublished photo, Col. Werner Baumbach, General of the Bombers, speaks at a Nazi Party rally in the Berlin Sports Palace. Noted Dr. Schmidt: "The code name for Speer's fantastic [1945 escape] project was—of all things—*Winnetou*. This was the name of a character in the Wild West novels of Germany's most popular writer, Karl May … who was condemned later for being Germany's greatest dilettante. Speer's plan was to join forces with Col. Werner Baumbach, a fighter pilot, and several other friends. The group would then fly a seaplane to Greenland, where they would wait for the end of the war and the occupation of Germany. Speer assumed he would return two months after the capitulation 'in order to take over the German government.' In his memoirs, Speer tells about his escape plans … but he conceals his political goals. He claims that he and his friends intended to fly to England in the autumn of 1945 and surrender there… On April 25, 1945, fighter pilot Baumbach and Armaments Minister Speer met 'in the forest camp at nightfall … '*Winnetou* plans rising,' Speer noted." When Hitler appointed Admiral Karl Dönitz Reich President, however, Dr. Speer decided to remain in northern Germany after all, and thus *Winnetou* was abandoned. (HHA)

Hess in his duties, if not his title, when the latter had flown to Scotland, and one of his operatives was a spy within Speer's own Ministry—Xaver Dorsch, Head of the Organization Todt, the construction arm of the Third Reich. Dorsch had been an admirer of Dr. Todt, and had had hopes of succeeding him before Hitler had named Speer.

For the 10 weeks that Speer lay in the hospital at Hohenlychen, Dorsch was the linchpin behind a secret cabal to overthrow him that not only included Bormann, but also Dr. Goebbels, Dr. Robert Ley (who wanted Speer's job outright), and Reich Marshal Hermann Göring, who had lost many of his previously held economic Four Year Plan powers to Speer in 1942. Indeed, within the Third Reich, Speer had developed a powerful host of enemies who were now determined to bring him down.

After the war, both Speer and Annemarie Kempf said they believed that the Reichsführer-SS was out to assassinate him medically, and the secretary even claimed to have overheard a conversation between Himmler and Dr. Gebhardt that concluded with the words to the doctor, "Well, then, he'll just be dead!" Himmler was already making inroads into Speer's domain.[53]

Dr. Gebhardt had previously participated in the medical treatment of SS General Reinhard Heydrich, who was wounded on May 27, 1942 in an assassination attempt in Prague. He was carried out of Prague's Bulovka Hospital in a coffin on June 4 after Dr. Gebhardt had been sent by the Reichsführer-SS to take over the case. Was Speer to be next? Gitta Sereny notes,

On admission, Gebhardt's clinical notes say, "the patient appeared exhausted. Exceptionally taut swelling of the left knee joint. We immobilize the leg and apply arnica poultices. Diet: vegetarian and fruit."

When there was no improvement after five days, he ordered massive doses of sulfa. Eight days after admission, although Speer showed general cold symptoms—bronchitis, hoarseness, and nasal catarrh—and although the consultant's registrar suspected pleurisy, Dr. Gebhardt stuck to his diagnosis of rheumatoid inflammation of the left knee.

Although a retrospective study of Dr. Gebhardt's clinical reports clearly establishes that he misdiagnosed his patient—who either already on arrival had the beginnings of an embolism or developed it in the course of that week—it is highly doubtful that, given Speer's determination to continue working, any physician could have done much better.[54]

Meanwhile—while the palace revolt went on within his Ministry of Armaments and Munitions at Berlin—Fraulein Kempf remained at his side constantly. When it appeared that Speer had taken a turn for the worse and might actually die, she called his wife, urging her to come, and also to get another doctor for a second opinion. She did so, bringing onto the case Professor Friedrich Koch.

Under his care, the crisis passed on the night of February 11–12, 1944, leaving the patient in what has been described as a drug-like trance. Speer himself later stated: "I've never been afraid of death since. I'm certain it will be wonderful."[55] Dr. Koch noted for the 15th: "An astonishing recovery … breathing normal, bleeding stopped… Leg and knee suddenly normal, no sign of phlebitis, no other physical symptoms."[56]

The origin of the inflammation of the knee and then the left lung remained "a mystery." Dr. Gebhardt had wanted to perform an operation to puncture the left lung, but Dr. Koch had declined. Speer also thought that the SS doctor wanted to poison him.

Speer was moved first to the grounds of Castle Klessheim in Austria, the German Foreign Office's guest facility for visiting heads of state, and it was there—after 10 weeks—that he saw his Führer for the first time since his illness began. Their reunion was a cold affair, with both men noticing the difference from former times when their being together as "fellow architects" had been so looked forward to. Now, Speer would recall later, he believed Hitler to be a criminal who was bringing death and destruction to Germany, and the end to all their joint building plans as well, not to mention the lost war and the Holocaust in the east for which the entire leadership corps of the Third Reich would one day have to pay with their necks. "I had begun to think about the course I was pursuing at his side, after 10 weeks of having not seen him… I no longer had a goal."[57] Speer was beginning to take his emotional leave of Hitler, and planning on how to survive both the fall of the Reich and the loss of the war. Allegedly, he began feeling a new responsibility; not to Hitler, but to the Reich itself.

From Salzburg, the entire Speer family left for a six-week recuperation stay at Goyen Castle near Merano, Italy where Speer decided to resign from his post. He submitted his resignation to Hitler

A German Alpenkorps (Alpine Corps) soldier carries a wartime *Panzerfaust* in this previously unpublished artwork. Both the hand-held Panzerfaust ("antitank fist") and the *Panzerschreck* ("terror of the tanks") were hastily developed after the Wehrmacht encountered the American GI "bazooka" in Tunisia in 1943. Author Eugene Davidson noted in his 1966 study, *The Trial of the Germans: Nuremberg 1945–46:* "Five million panzerfausts were produced in 1944 alone: one weapon for one man." Adds van der Vat: "3½ million were made from November 1944 on," but, "They were no match for accurate anti-tank guns." British author Max Hastings says in his work *Overlord:* "The German Panzerfaust was the best hand-held infantry antitank weapon of the war, exceptionally useful in Normandy where the close country made it possible for its operators to reach Allied tanks at the very short ranges for which it was designed. It was a one-shot, throwaway weapon, weighing 11½lb. Its hollow charge could penetrate 200mm of armor at 30 yards, and an improved version—introduced in the summer of 1944—possessed a higher velocity, and was effective up to 80 yards." (US Army Combat Art Collection)

Above: Hitler (center) silently ponders new German small arms weaponry during a wartime demonstration. Just over his right shoulder can be seen SS Gen. Karl Wolff, and at the far left is Speer's left side, just inside the frame. Stated one source, "The Wehrmacht would in future receive top priority in the fulfillment of orders. Speer's Ministry took over the control of Wehrmacht orders, and determined which companies were to carry them out, and the production method to be used. Speer's organizational plan was an enormous success. Arms and munitions production increased by about 60% in the first months (March–July 1942) of his leadership." (Previously unpublished photo, HHA)

Right: A famous wartime photo of a King Tiger tank carrying paratroopers during the Battle of the Bulge. Dan van der Vat attests that Speer selected tanks as his personal area of expertise, even though he had "Panzer" Röhland as his personal advisor in this field, but especially in that it was the one in which Hitler himself was considered to be expert. (CER)

on April 19, 1944. While Göring fumed that he simply could not do this, Hitler raged to his secretary Johanna Wolf that it was "impertinent." At Merano, Speer was "guarded" by 25 SS men.

The following day, Hitler's birthday, Speer received two visitors. The first was industrialist Walter "Panzer" Röhland, who begged Speer to remain at his post, using for the first time the words "scorched earth" that Soviet dictator Josef Stalin had used to halt the German drive outside Moscow in 1941 and that had so impressed Hitler at the time. Would the Führer employ the same methods in regards to the Reich? Röhland believed that he would, and for this reason alone, Speer must remain at his post, he asserted.

A delegation headed by Speer's ally Luftwaffe Field Marshal Erhard Milch also arrived unexpectedly to plead with Speer not to resign, and to reassure him that he still retained Hitler's favor. An enraged Speer blurted out, "The Führer can kiss my ass!" to which the shocked Milch replied, "You are much too insignificant to use such language toward the Führer!" in an attempt to cut him down to size.[58]

Speer decided to reconsider his position. Meanwhile, Dr. Gebhardt had told everyone that Speer was incapable of returning to work, Hitler told Frau Speer he might die (as Göring also intimated to the patient), and the Reich Marshal was also gleefully shopping around for a successor.

Later, Speer decided to fly to see Hitler at The Berghof (Mountain Home), in Bavaria. Dr. Koch approved the flight on medical grounds, but Dr. Gebhardt balked. Dr. Koch recalled later: "He again accused me of not being a 'political doctor.' Here, as in Hohenlychen, I had the impression that Gebhardt wanted to keep Speer in his clutches."[59]

At The Berghof, Speer was received by the Führer as a visiting head of state: "Hitler had donned his uniform cap and, gloves in hand, posted himself officially at the entrance... He conducted me into his salon like a formal guest... Although the old magic still had its potency, although Hitler continued to prove his instinct for handling people, it became

Germany Navy Grand Adm. Dr. Erich Raeder (1876–1960), seen here in a formal portrait with ornate baton of rank, was a colleauge of both Drs. Todt and Speer. (HHA)

increasingly hard for me to remain unconditionally loyal to him."[60]

Nevertheless, whatever differences there were between the two men were papered over—at least for the next year. Dorsch was restrained and placed once more under Speer's complete control. Martin Bormann was defeated on this and other issues as well, and tried unsuccessfully to cultivate a friendship with Speer. Göring retreated back to his hunting preserve at Karinhall, Dr. Ley's plan to succeed him was aborted, and Dr. Goebbels realigned himself with Speer in time for the July 20, 1944 plot. Indeed, that very day, the two men were together.

Oddly, a year later, Speer recommended Dr. Gebhardt to his friend

Jean Bichelonne, the French Minister of Production, for an operation; he died. Moreover, for Speer himself the danger was not yet over, as his own subordinate, Walter Brugmann, died in a mysterious plane crash on May 26, 1944, very similar to that of Speer's own predecessor, Dr. Todt. Had someone sent him yet another warning of his mortality?

As Speer noted in his memoirs, his absolute loyalty to the Führer and the Nazi Party had been shaken by these events: "I had learned the valuable lesson that a resolute stand with Hitler could achieve results [in suppressing the Dorsch revolt]… I was beginning to bid farewell."[61]

He returned to work. Among his top men at this time were industrial director C. Krauch, Reich Commissioner for Coal Paul Plieger, Leuna Chemical Works chief Butefisch, I. G. Farben Chairman Fischer, Chief of Planning and Raw Materials Dr. Kehrl, munitions organization head Edmund Gelienburg, head of the power industry Dr. Carl, Dieter Stahl,

Grand Admiral Dönitz (second from left) shakes hands with a German naval rating (right) during an inspection trip, his staff officers behind him. During the war, he and Dr. Speer had an excellent working relationship, one that disappeared at the 1945–46 IMT trial, however, never to be regained. (CER, LC)

in charge of ammunition production; and Dr. Friedrich Luschen, head of the Reich electricity industry. Speer recalled: "The industrialists were made of sterner stuff than Hitler's entourage. They held fast to their verdicts, supporting them by data and comparative figures… The failure of the Ardennes Offensive meant that the war was over."[62]

In his speech before the Central Planning Board of March 1, 1944, Minister Speer asserted: "Out of five million foreign workers who arrived in Germany, not even 200,000 came voluntarily." He stated that there

SPEER IN THE BRITISH PRESS

Several contemporary British newspapers published feature articles on Dr. Speer that are of interest because they convey period views of him as seen by his enemy. This was published in *The Times* on September 7, 1942.

THE SPEER PLAN IN ACTION: MOBILIZING RESERVES OF MEN AND MATERIAL
Hitler's New Drive for Victory
From Our Special Correspondent on the German Frontier

Every feature of the vast totalitarian drive now in progress on the German home front supports the tacit assumption now common in the Third Reich that the Axis must at all costs win the war this year. The decree issued in July by the Reich Minister for Armaments and Munitions, Professor Albert Speer, inaugurated "The mobilization of all remaining latent reserves of iron" to broaden the base of steel production required for the German war effort.

The totalitarian character of this effort to muster the last remaining reserves is apparent from the fact that not only old iron, scrap bars, and sheets of iron and steel in sizes seldom wanted, finished parts in iron and steel made in execution of orders since cancelled, iron and steel spares rendered obsolete by change of types, but even modern machinery and whole industrial plants in so far as they are not adaptable for immediate use in armaments production are to be scrapped without compunction. "Iron and steel delivered in the course of this action will in general be paid for at prevailing market prices for scrap." Valuable industrial plant in perfect condition, idle only because of the war will be destroyed and its owners dispossessed. That this constitutes the destruction of immense capital values and heavy loss of industrial potential is not denied. On the contrary, it is declared officially that, "All usual financial considerations and regard for postwar production are to be left out of account."

Hitler's Associate

This drive to mobilize all recoverable iron and steel is but part of a still vaster action now known in Germany as the "Speer Plan," designed to serve as the blueprint for the German war effort till the end of the conflict. "Professor" Albert Speer—engineer and architect by profession, now 37 years of age—is one of Hitler's most intimate associates. He joined the SA in the early days of the Nazi movement, where he became a close friend of Heinrich Hoffmann who, after the advent of Hitler to power, was made "professor" and "court photographer."

It is related that—night after night—Hitler and Speer used to sit till daylight in the garden behind the Old Reich Chancellery in the Wilhelmsstrasse planning the rebuilding of the Third Reich. Berlin, Munich, Cologne, Hamburg, Vienna, and every other leading German city were to be transformed. Speer was entrusted with the building of the New Reich Chancellery (NRC) and the re-planning of Berlin. The "Ost-West Achse/East-West Axis—completed just before the war—is generally regarded as a successful piece of work.

Speer was extremely ambitious. He was also jealous of the fame of the late Fritz Todt—builder of the Reich motor roads—the preliminary step towards the motorization of road traffic in the Reich with possible extensions over the whole Continent. None but a master like Todt could have so successfully improvised the solution of the advance of the German Army into Russia last year.

When, in October 1941, the German High Command realized that the campaign in Russia would tax the strength of the Reich far beyond anything at first contemplated—especially after it seemed certain that within a measurable period, Americans would join in the conflict—a plan for mobilizing all latent reserves of men and materials was hastily drawn up by Dr. Todt and General Fromm.

A Clash of Views

Todt and Fromm differed on certain points, but both were convinced that—as the war was not likely to end soon—it was imperative to conserve staying power and keep industrial potential unimpaired: hence, that it would be fatal to attempt to augment manpower at the front by withdrawing men and materials from war industry. Speer—whose views at that time did not carry decisive weight—held a different opinion. Apprehensive about the immensity of Russia and the impact of new armies from America, Speer urged that the war must be ended in the shortest possible space of time. Otherwise, not only would Germany lose the war, but the Nazi Party would be swept out of existence.

His view won the support of Hitler and the Party. Speer drew up his own plan for bringing the war to a rapid close. Todt rejected the plan. For the moment, it was forgotten. Immediately after Todt's mysterious death in an airplane accident near Smolensk, Hitler appointed Speer as his successor. That implied the adoption of the "Speer Plan."

Speer declared: "We must win this war by the end of October, before the Russian winter begins, or we shall have lost it once and for all. Consequently, we can only win with the weapons we have now, not with those we are going to have next year. We can only win if every man capable of bearing arms is at the front, and we must put all we have in the front line, whether that is going to be deleterious to postwar industry or not."

Speer was able to persuade Führer and Party because of his privileged position, which gave him access to Hitler at all times.

Speer used a term coined by [Walter] Rathenau in 1918 to explain a similar policy to meet a similar situation, 'levée en masse'/total draft mobilization of all available men and materials. Speer intended this to take place at the end of May, but—owing to opposition from the Wehrmacht, and from industry—discussion languished. About the middle of June, however, industrial sources revealed that workers and employees were being withdrawn in all masses from works, factories, business, and administrative bodies, and sent to the front.

Use of Foreigners

About the same time, the new State Secretary for Labor Allocation—Gauleiter Sauckel, appointed early in April—began drastic recruiting outside the Reich in France, Belgium, Holland, and Norway to obtain skilled industrial workers to take the place of those withdrawn from German factories by the Speer Plan. Sauckel went so far as to comb out again the so-called "Uberkompanien" fortifications of ex-servicemen so badly disabled as to be presumed unfit for any kind of military use, even after complete recuperation. Sauckel also turned the searchlight on all offices of the Wehrmacht and Party to occupied countries, sending to the front every able-bodied man... [Pierre] Laval's attempt to send 350,000 French skilled workers to take jobs in Germany was an essential part of the plan.

Even the figure at which Laval first aimed is highly significant. According to preliminary estimates, the total draft was expected to yield another 400,000 or 500,000 men for the front. To fill the labor vacuum in German industry, Laval was instructed to take 350,000 skilled workers in France, the expectation being that the other 50,000 to 100,000 skilled workers would be found somehow in other occupied countries, although these had already been drained. Actually, probably because the comb-out was more drastic than intended, the yield in Germany, according to

statements by industrial and official quarters, was "about 4% of the total working population." [4% of the total workers would be somewhat over 400,000; 4% of the whole working population would be 800,000. Probably the net result is somewhere between these two limits.]

Laval's Fiasco

In any case, the withdrawal of so many skilled workers from German industry could not but cause prolonged dislocation. Hence, the rancor felt in Germany against Laval, whose efforts to obtain substitute labor has been such a fiasco. Berlin makes the additional reproach that most of the workers who volunteered to go to Germany are not French at all, but foreigners naturalized in France only a few years before the war.

Against this background, the mobilization in Germany of the last iron reserves is fraught with meaning. Estimates of the amount of iron and steel obtainable from the drive speak of "many million tons." What this means is not certain. Speer has, for instance, mentioned that a great deal can be recovered in practice by postponing repairs and renewals. At least theoretically, the use of iron and steel obtained by the methods prescribed for the drive are certain to effect substantial, though temporary, economy in coal, transport, and metal.

Scrap yields about 85% its own weight of steel. A high percentage of the coal used on smelting ore is saved. Moreover, the transport of scrap to the blast furnaces is more economical than the transport of ore. How important the saving in transport in is evident from the fact that it is admitted that booty taken on the Russian Front cannot at present be used for steel making because it cannot be transported to the blast furnace. Scrap collected in Germany can be conveyed on empty trucks as ballast by waterways. House building has been suspended to economize consumption of iron and steel, even though the destruction caused by the RAF makes new building an urgent necessity.

The drive can be made only once, however, so that the advantages are temporary, Speer himself has decided: "The maintenance of our present superiority in armaments over the enemy will depend to a large extent on the scrap we collect; not only our quantitative superiority, but also our effectual power... We must increase our output stage by stage as the output of our economy increases."

When there is no more scrap to be collected, the United Nations will begin to overtake the Reich.

were 400,000 POWs, with Russian prisoners being used in coalmines. "I was running a country's war machine, not a church bazaar," and was responsible, overall, for 28 million workers, according to Sereny. "I didn't see or think of them as human beings, or as individuals."[63] When asked by Norden about forced labor, in 1971, he said:

> ...despite my initial reservations, I was Sauckel's wholehearted collaborator on the forced labor program, and I share his guilt... Our roles were rather like the captain of the slave ship and the slave owner who buys his cargo... At that time, I was deeply grateful to Sauckel for each worker he sent me. When it appeared to me that we might fall behind schedule in our armaments production, I pressed Sauckel to provide me with more workers; and when we failed to fulfill quotas, I often shifted the blame to him.
>
> When Hitler instructed Himmler to provide concentration camp inmates for the plants, I welcomed them, as I did Russian POWs. I treated these millions of people as no more than servomechanisms for

Wrecked docks and other port facilities at Wilhelmshaven, Germany, 1945. The vaunted "fortress without a roof," Hitler's so-called Festung Europa (Fortress Europe) was the target of massive Allied bomber fleets night and day. By day, it was the Americans, and at night the RAF. The raid on Hamburg, August 17–18, 1943, has been called "One of history's most decisive air attacks," and Speer noted that "The attack on Hamburg put the fear of God in me." In 1944 Allied air attacks were concentrated on the systematic bombardment of key parts of the German armaments industry, in particular petrol refineries and oilfields. The gigantic Krupp steel plant in Essen in the Ruhr would be pounded by 55 RAF

bombing raids during the war. German gasoline production was already more than halved by the end of May 1944, and Allied attacks in June and July dealt a death blow. Production decreased by 98% and reserves had to be tapped. Despite these savage and extensive raids, German wartime armaments production actually increased as the war's crescendo peaked, much to Speer's credit, and also to the postwar astonishment of the US Strategic Bombing Survey. However, Professor Speer considered the air war a far worse military disaster for Germany than did either Hitler or the discredited Luftwaffe commander, Göring. (US Army Signal Corps)

our machines of war, and that is both a legal and a moral crime…[64]

In April 1944, command of the OT was taken away from Speer by Hitler and given instead to his deputy, Xaver Dorsch, because Speer had not acted upon Hitler's request for "concrete factories" for certain products—crankshafts, gears, electric devices— given in the spring of 1943. Dorsch took over aircraft production on March 5, 1944. Saur was given Fighter Plane Staff construction command as well. Speer's star was waning with Hitler after a heady two-year run.

In Speer's view, his decline at least in part was due to "The old animosity of high-level Party functionaries, who always saw me as an upstart in their ranks, a man who had bewitched Hitler with his architecture, and thereby sneaked illegitimately into the close circle."[65]

Speer recalls that, "The technological war was decided," by the pre-D-Day Allied bombing raids of May 1944 by 1,935 bombers of the USAAF. "A new era in the air war began. It meant the end of German arms production."[66] Despite the bombing, however, enough fuel had been stockpiled to carry on for the next 18 months, or until December 1945, by his calculations.

It wasn't until September 1, 1944—at the beginning of the fifth year of the war!—that Speer convinced Hitler, Göring, and Sauckel that German war production *must* draft all available German women into service.[67] In his 1971 interview, he said:

> Because of our manpower shortages, I recognized the prime importance of employing German women in industry. This had been done with considerable effect in the First World War, and was already standard practice in the Allied nations, but Hitler would not hear of it, nor would the officer responsible for manpower, Gauleiter Fritz Sauckel.
>
> This was a crucial error. If, from the beginning, we had followed the Allies' lead, we could have had over one million German women working in our factories, freeing one million men for the Army, but the refusal to entertain my proposal had a more sinister result. It was a major contributing factor to our program of forced labor from the occupied territories—which led to my own conviction at Nuremberg, and 20 years in prison.[68]

There was a reluctant response to the decree that conscripted women for war work in factories. To help women overcome their apprehensions about war work in factories, plant owners were encouraged to adopt a number of innovations. A special agency was created to see that factories were made reasonably comfortable and safe. By law, women were exempt from carrying heavy loads. Some factories planted gardens where women could relax during breaks. Many others introduced training programs in which experienced women workers helped newcomers learn their jobs. Also, although it was common for workers to do 56 hours a week in the German war

SPEER AND THE SS

In the spring of 1944, Reichsführer-SS Heinrich Himmler achieved a goal first set out a decade earlier by the ill-fated SA Chief of Staff, Captain Ernst Röhm: the independence of his paramilitary organization from both Party and State, thus establishing the basis for a postwar civil war between it and the German Armed Forces had the Third Reich won World War II. Part of the SS's planned financial independence was the establishment of an internal industrial empire to rival both the Hermann Göring Works and also Speer's Ministry of Armaments. This was but another step on the long path toward Himmler's eventual goal of an SS state *within* the overall Nazi state.

As Speer noted in *Infiltration*, the SS failed in its wartime industrialization plot because it could not make industrialists out of devoted but ignorant—SS members. When Himmler visited SS camp commandant Rudolf Höss at Auschwitz in March 1941, he told the latter that it would be an armaments manufacturing facility with 100,000 workers. It is my conclusion that a major reason behind Dr. Speer's writing *Infiltration* was to once again shift the blame from himself to the SS in regards to the afflictions of the Jews.

The original plan of March 16, 1942—in the very first month of the Speer ministry—was to employ 25,000 concentration camp laborers in five SS camps: Buchenwald, Sachsenhausen, Neuengamme, Auschwitz, and Ravensbrück—and it all grew from there. By September, Speer noted, "50,000 Jews would soon be occupied in existing factories."

Speer, however, reversed the main SS process: instead of factories being built outside existing SS camps, the camps were built near the operational factories for the next 2½ years, September 1942–December 1944. The pecking order in the relationship between the Reichsführer-SS and Dr. Speer was that—as the senior Nazi—Himmler made Speer report to *him* for conferences, and never visited the Minister of Armaments in *his* home or office, except once, in March 1945. Speer termed Himmler "An utterly insignificant personality" nonetheless, perhaps a swipe at his late rival.

Despite the vaunted SS efficiency, Speer noted that by March 1943, they had delivered but a thousand concentration camp inmates to work at Buchenwald, not 5,000, "due to the lack of ability in the SS agencies." There was also SS bribery and corruption in the usage of wartime materials.

When I compare the total sum of our production to the SS performance, it is easy to see how minimal their overall contribution to armaments was. In this period of seven months, the SS did not contribute more than two thousands of a percent to the overall armaments output... Never did the armaments production of the SS reach the scope attributed to it by Himmler's imagination.

A wartime *Waffen/Armed* SS recruiting poster that linked the heroic Germanic Siegfried warriors with the planes, guns, and tanks provided to them by the Speer Ministry. By the end of the war, the SS had fully infiltrated both Speer's offices in Berlin and the overall German arms industry as well. (LC)

Furthermore, the SS factories in concentration camps were extremely unprofitable because of the cost of guards and barracks.

In April 1943, the SS concentration camps housed 171,000 prisoners.[69]

By January 5, 1945, "there were 714,211 in the Reich alone." Himmler, Dr. Goebbels, and Minister of Justice Otto Thierack agreed to the policy of "annihilate through work" the Czechs, Poles, and gypsies in the areas of German occupation.

On August 20, 1943, Himmler achieved another of his career goals: the ouster of his nominal superior—Minister of the Interior Dr. Wilhelm Frick—and his own assumption of that post by Hitler's decree. When the new Reich Minister entered the Führer's military conference at FHQ, Hitler looked at him, and then Speer, and referred to them as "You two peers," thus making public the fact that each of them was now on par with the other; in effect, maximum rivals for power. By this time, too— as Speer noted—the Gestapo "had its agents in almost every office and factory" within the Ministry of Armaments' area of responsibility. The SS also had its own Ordnance Office, like the Army's.

In the period 1942–43, Dr. Speer had relatively smooth sailing with the SS, but in the first six months of 1944, the Reichsführer-SS tried to overthrow him as a dangerous rival, perhaps as a result of Hitler's deliberate and considered, "You two peers" comment of the previous year. In the spring of 1944, Speer had his personnel department sealed safe opened, and discovered from the documents found there that Himmler had indeed infiltrated his ministry with several spies. Speer's main nemesis within Himmler's large apparatus was the Reich Main Security Office headed first by Reinhard Heydrich, then by Himmler, and finally by Ernst

SS Lt. Gen. Oswald Pohl (center), the former Chief of the SS Economic and Administrative Office, was discovered working as a casual laborer in the spring of 1947. Here, he is seen with a US Army Military Policeman (left) and an interrogator (right). It was Pohl's office that was responsible for the running of the SS concentration camp

empire from 1942 that was used by Dr. Speer. In his memoirs, Speer described Pohl as "tough and ruthless in negotiations." (US Army Signal Corps photo by Ray D'Addario)

Kaltenbrunner. It was this agency that in the final nine months of the war was seen as the last and ultimately most dynamic Third Reich organization. Indeed, fully 77 members of the OT were also in the SS, as was he himself.

"My power was broken," Speer wrote of the successful attempts of the Reichsführer-SS in October 1944 that removed a trio of his major deputies, and thus left him seeking new roles: first, to prevent Hitler's destruction of Germany proper; second, to become chief—under the Allies, if necessary—of the Fourth Reich's postwar production. As for Himmler, by June 1944, "He intended to develop SS-owned steel production all the way to the finished product."

By March 1944, Speer also found himself personally "guarded" by the SS, as he recalled in 1981. Up until then, "I was not one of the functionaries who were accompanied night and day by a unit of some four–six SS men… These included Dönitz, Keitel, Göring, Goebbels, Rosenberg, Funk, and other ministers." Thus, not only could Hitler order their arrest at any time—as he did with Göring in April 1945—but so could Himmler. By 1945, Speer somehow eluded these guards in order to make his many automobile drives across the Reich undoing Hitler's "Nero Orders" to destroy everything.

By March 1945, the prisoners working in SS camps were guarded by over three divisions: 36,454 men—troops that might better have been deployed at the fighting fronts. Nor was the Jewish labor employed in the camps cost-efficient, Speer noted. They "performed only a fraction of the labor that they had previously performed in independent factories," under German civilian industrial supervision. They made "20 million cores for infantry munitions," he remembered.

Ironically, Dr. Speer's own man in the SS-occupied General Government in Poland was none other than SS Lt. Gen. Oskar Schindler, later made famous by the 1994 film *Schindler's List*. Indeed, even the duo of Himmler and his minion Oswald Pohl found themselves at odds with their own subordinate, Rudolf Höss, and this central command dichotomy mirrored the schizoid dual simultaneous SS policies of industry usage versus extermination of the Jews.

The postwar Himmler-Kammler vision of an SS State encompassed Scandinavia, The Netherlands, Poland, Czechoslovakia, Burgundy, Alsace-Lorraine, Belgium, the Austrian Tyrol, and the French coal district near Lille, plus Himmler's planned colonial "German India" settlements as far as the Ural Mountains, the Volga River, and the Baku oilfields on the Caspian Sea. These plans undoubtedly would have clashed with both Göring's Four Year Plan and Rosenberg's Ministry for the Occupied Eastern Territories—just as Hitler planned for under his "divide and conquer" rule. Which would have survived a postwar Nazi–Army civil war is difficult to say, as is Speer's place in all of this. My guess is that he would have sided with Göring against a probable Himmler–Rosenberg alliance, and that the Army would most likely have joined the Speer–Göring combine, but this is all empty speculation.

industry, some employers instituted shorter shifts for married women. The government also expanded childcare facilities; by 1944 there were 32,000 nurseries catering for the 1.2 million extra children needing care while their mothers worked. For practical reasons, these working woman began to wear trousers, much to the dismay of many German men. One Army officer at Garmisch-Partenkirchen in Bavaria forbade his men to be seen in the company of "trouser women." This order was overruled by Goebbels, who said, "Whether women wear trousers is no concern of the public... The bigotry bug should be wiped out." Goebbels must have recognized that such old-fashioned scruples could not be allowed to continue if they affected the war effort.

From fall 1944 onward, Speer was perceived by Hitler—like Field Marshal Erwin Rommel before him—as being a defeatist, while Technical Office Chief Saur's political stock was rising.

"I had introduced a fairly successful 'Americanism' into the armaments organization, and this innovation was a decisive step toward the managerial revolution of German industry," Speer asserted. The "optimal factor" was in achieving the highest performance with

A previously unpublished photo of Dr. Walther Funk (left), Minister for Economic Affairs, congratulating Hitler (right) after his having survived the July 20, 1944 Army bomb plot to kill him. At left center stands Field Marshal Wilhelm Keitel. (HHA)

Speer's wartime colleague, Luftwaffe General of the Fighters and ace Adolf Galland, wearing the Knight's Cross of the Iron Cross with Oak Leaves at the throat. (HHA)

Dr. Speer's wartime colleague, Luftwaffe General of the Bombers Col. Werner Baumbach (left), being awarded a medal by Hitler at FHQ Vinnitsa in occupied Soviet Ukraine, as seen on the cover of the September 10, 1942 issue of the *Illustrated Observer*. (RHA)

the least amount of manpower and material. The fatal flaw, however, was that German higher quality tanks, weapons, and jets "Always came too late…" Basic production of weapons in the Ruhr had stopped in November 1944, and from then on "German industry could only assemble weapons from the remaining stocks of individual components…"[70] As of December, 1944, the Arms Ministry had collapsed for good.

Minister Speer made three Western Front inspection trips late in 1944, including a flight to the Ruhr on November 15, "Visiting Krupp in Essen, Dortmund, Bochum, the fuel plants at Scholven and Geisenberg, the power plant at Goldenberg, nitrogen and piping plants near Cologne and Düsseldorf, and bomb-damaged areas nearby," noted Sereny.[71] His "Eastern trip" encompassed Czechoslovakia, Hungary, Danzig, and Silesia.

On December 1, 1944, Minister Speer gave his last speech to the heads of the arms commissions and his ministry.

In January 1945—Professor Speer told Goebbels that even with all the occupied territories lost, the Reich could still produce arms for another year, until January 1946. This estimate had to be revised, though, when Upper Silesia was lost to the Red Army.

On March 18, 1945—in the underground Berlin Führerbunker built by Speer, Hitler gave his infamous "Nero Order," aping that of Stalin before the Battle of Moscow in 1941: that everything of any value was to be destroyed in the path of the advancing Allied armies as they invaded Germany. This was to include power plants, bridges, electrical stations, crops, cattle, waterworks, and anything else of possible use to the enemy. "We can no longer concern ourselves with the population," Hitler observed to his intimates, and "No one objected," Speer noted.[72]

The next day—on his 40th birthday—Minister Speer presented to Hitler the later-famous 22-page memo he had written on March 15, "In which I dryly set forth the collapse of his whole mission." It opened with the words, "The war is lost." Hitler rejected it on March 27, asserting that, "If the war is lost, the people are also lost."[73]

On March 23, 1945, Minister Speer and his adjutant Manfred Poser drove back to the Ruhr. Speer ignored Hitler's "Nero Order" to destroy everything, and visited his parents at Heidelberg for the last time. "I think they loved me then," he recalled.[74] His father died in April 1947, and his mother in June 1952, while he was imprisoned.

According to his military aide Manfred Poser, Minister Speer made 18 trips and held 35 meetings in February 1945; 31 trips and 35 meetings in March; and 19 trips and 28 meetings in April: "Thousands of kilometers by car, and also plane."[75] He tried to persuade the field commanders to ignore Hitler's order, and not implement a "scorched earth" policy.

Speer returned to Berlin from his Western Front trips on March 29, 1945 at 1a.m. The Ministry of Armaments had virtually shut down earlier that same month, and thus there was nothing more there to do. In Berlin he was summoned to the Chancellery to face Hitler. When asked, he repeated his assertion that the war was lost. In a tense confrontation, Speer was ordered on sick leave by the Führer: in effect, being fired. Speer objected, stating that he wasn't sick, and therefore would only resign his post. Hitler said he could not accept his resignation at such a time, and requested that he take 24 hours to decide whether he could at least have hope in victory. Speer returned the next day, asserting, "My Führer! I stand unreservedly behind you." Then—according to Speer—a grateful Hitler observed, "Then all is well," and Minister Speer was allowed to keep his post. "Thus I was reappointed."[76]

Minister Speer returned to sabotaging Hitler's orders to destroy Nazi Germany from within, which, in my view, Hitler probably knew all about and of which he may have secretly approved.

According to Gitta Sereny, however, Speer's critics believed instead that he cracked under Hitler's ultimatum. Still, the team of Speer and Gen. Gotthard Heinrici managed to save from planned demolition 866 of Berlin's 950 bridges. The US Army left the Ruhr's three industrial cities intact. Walter "Panzer" Rohland radioed Professor Speer from the Ruhr: "We'll always be in your debt."[77]

Meanwhile, Saur moved what remained of the Ministry of Armaments to the Harz Mountains, and Hitler—isolated in Berlin—divided Nazi Germany in half geographically, with Grand Admiral Dönitz in charge in the north, and Luftwaffe Field Marshal Albert Kesselring in the south. Speer sent his family north to the Baltic Sea area for safety. When Franklin D. Roosevelt died on April 12, 1945, Dr. Goebbels wanted Hitler to send Speer to see the new US President, Harry S. Truman, as a negotiator. Hitler rejected the idea.

On April 20, 1945, the top Nazi satraps gathered in the Berlin Führerbunker to celebrate Hitler's 56th—and last—birthday. At the party, Hitler announced that plans had changed, and Berlin would be defended to the last. Goebbels endorsed the decision, saying he wanted to die in Berlin, his home. Others started to make excuses to leave Berlin. Speer was at the party, and then left Berlin for Hamburg. Dr. Hupfauer recalled how "Speer was sure he'd be needed by the Allies—especially the Americans."[78] Manfred Poser agreed. As for Minister Speer, he watched the Battle of the Oder against the Red Army that the Germans lost. The Russians used "Todt's beautiful autobahn" in their steady advance on Berlin in April 1945.

In advance of the Red storm, Speer had Hamburg declared an open city—and then decided to fly back to Berlin to say goodbye to Hitler. Annemarie Kempf called this a "death wish," but critics believed it was to ensure that he wasn't appointed by the Führer to be his successor. That would negate his plans for a position under the Allies. Instead, they charged—and Dönitz believed—that Speer engineered the appointment of the Grand Admiral as the new Reich President (using Paul von Hindenburg's former title).

"On April 23rd, I flew to Berlin to see Hitler again."[79] He wasn't alone, however, but had a fighter escort of a dozen planes on his flight in a small Storch courier plane from the Luftwaffe's test station at Rechlin to the capital's Gatow Airport. At this point, the Russians were only blocks away.[80] Discussions as to whether to stay or leave were ongoing among those remaining in the bunker. Having said goodbye to Hitler, Speer flew out again the next day. It is thus possible that Speer asked Hitler to exclude him, and that the Führer granted his wish.

In his Last Will and Testament, Hitler himself dismissed Minister Speer in the new Dönitz Cabinet, replacing him with his arch-rival, Karl Otto Saur instead. Speer cattily described his successor as a coward. Hitler's final secretary—Traudl Humps Junge—noted in her memoirs that Minister Speer's final departure from the Bunker was "hardly noticed," and that afterward, "Hitler did not talk about Speer any more."[81]

Some of the Focke-Wulf Fw 190s abandoned to their fate as the war drew to a close. By 1945 Drs. Todt and Speer had overseen the production of 13,367 Fw 190A fighters in 10 different models, as well as an additional 6,534 Fw 190 fighter-bombers. Its armament consisted of four 20mm cannon and a pair of 7.92mm machine guns. (Smithsonian National Air and Space Museum, Smithsonian Institutions, Washington, DC)

Speer flew on to Hamburg, but when Red Air Force fighters were spotted, he went to his own headquarters at Eutin, located in some trailers on Lake Eutin, with General of the Bombers Col. Werner Baumbach as his pilot. There he gathered about him his final staff at HQ Speer. Then he flew by Storch again to see his own family one last time in their Baltic coast eyrie.

Back at Eutin on May 1, Speer heard of Hitler's suicide the day before—and then left to join Reich President Dönitz at Plön. The new government left there the next day for the larger—and safer—naval base at Flensburg, where the Grand Admiral appointed Dr. Speer his new Minister of Industry and Production. Speer drove daily to the 10a.m. Cabinet meetings, discussing "Non-existent plans for a non-existent country."[82]

While he was at Flensburg, he was "interrogated" by John Kenneth Galbraith and George W. Ball, directors of the US Strategic Bombing Survey, about the effects of Allied bombing. They later published an article about the seven days they spent interviewing Speer. Their article includes a good summary of what Speer felt were his achievements as minister:

The German High Command on Hitler's 53rd birthday, April 20, 1942 at FHQ Wolf's Lair, Rastenburg, East Prussia. From left to right: unknown officer, with back to camera; Field Marshal Keitel shields his eyes from the sun, unidentified Foreign Office aide, Gen. Walter Buhle, Karl Otto Saur (1902–66), Hitler, Reich Marshal Göring, three unknown officers, Minister Speer, Grand Adm. Dr. Erich

German aircraft production was greater a month or two after the industry was bombed than it was before. Speer considered this his proudest achievement, but he was also proud of the way ball-bearings and tank plants had been restored to near full production within a few months after they were hit. He did not succeed in restoring synthetic oil plants under repeated bombing, but he believed he had made a creditable effort—more than 330,000 men had been put to work rebuilding the devastated plants. In succeeding days, he told how he had moved in on one stodgy German industry after another to force reorganization of production methods and increased output. He also told how he forced the German Army to simplify its weapons and eliminate elaborate hand finishing to adapt them to mass production. These were not empty boasts. Under Speer's management, German war production increased threefold.

Speer was especially anxious that we should not suppose that such things were easy in a dictatorship. He referred repeatedly to his political difficulties. Bormann—as chief Nazi politician—tried to make him hire Party hacks for responsible positions, and harassed the non-Party technicians that Speer employed. Speer eventually had to fire his ace trouble-shooter—one Schieber—because Bormann considered him

Raeder, SS Gen. Karl Wolff, Dr. Robert Ley, unknown Luftwaffe officer, and Reich Chancellery chief Dr. Hans Heinrich Lammers. Of Saur, Domarus noted: "Hitler no longer called on Speer in this connection," i.e, as "Head of the Technical Office in the Reich Ministry for Armaments and the military-industrial complex." (HHA, US National Archives, College Park, MD)

politically unreliable. Sauckel—the manpower tsar—was another source of trouble. He refused to draft German women for war jobs, would not supply labor where and when Speer wanted it, and maliciously upset Speer's plans for reorganizing war production in France by making a special draft of workers from precisely those plants to which Speer gave contracts. Speer also accused the Army of drafting technicians who were urgently needed by industry, and wasting manpower in useless rear echelon jobs. Göring and Goebbels he accused of having no brains whatever. In the early years of the war, Speer's chief rival for power over German industry was Walther Funk, head of the economic ministry. Speer told of the troubles that resulted from this division of authority. He had finally settled matters by taking over all of Funk's responsibilities. This he managed partly by capturing the loyalty of the one subordinate on whom Funk (a notoriously lazy and bibulous man) depended.[83]

FROM CAPTURE TO NUREMBERG, 1945–47

"We had gambled, all of us, and lost."
Albert Speer

Speer at Nuremberg. His indictment read: "The defendant Speer between 1932–45 was a member of the Nazi Party, Reichsleiter, member of the Reichstag, Reich Minister for Armaments and Munitions, Chief of the Organization Todt, General Plenipotentiary for Armaments in the office of the Four Year Plan, and Chairman of the Armaments Council. The defendant Speer used the foregoing influence in such a manner that he participated in the military and economic planning and preparation of the Nazi conspirators for wars of aggression and wars in violation of international treaties, agreements, and assurances set out in Count One and Two; and he authorized, directed, and participated in the war crimes set forth in Count Four, including more particularly the abuse and exploitation of human beings in the conduct of aggressive war." (Previously unpublished photograph by IMT US Army Signal Corps photographer Ray D'Addario, Holyoke, MA)

On May 2, 1945, new Reich President Grand Adm. Karl Dönitz, as Hitler's successor, formed a new Acting Reich Government with former Nazi Finance Minister Schwerin von Krosigk as Principal Minister, not as Reich Chancellor. Professor Speer was named as well, to the new post of Minister for Industry and Production, with his former Third Reich associate, Minister of Traffic Dr. Julius Dorpmüller, as the new Minister of Post and Traffic. Dr. Herbert Backe remained as Minister for Food, Agriculture, and Forests. The Nazi salute was replaced by the former, traditional military salute, hand to cap.

The new Reich President also fired several former Nazi ministers, among them the Reich Minister for Science and Education, Dr. Bernhard Rust. Himmler was finally fired from all his posts, too, by Reich President Dönitz on May 6, 1945. He set out across northern Germany as a fugitive, fleeing Allied capture and trial as an accused war criminal. Overnight, both Nazi Party and, especially, SS uniforms disappeared from public view, with Dr. Backe changing to civilian clothes, as did Dr. Speer, abandoning his brown Organization Todt garb after more than four years.

On May 3, 1945, Reich Minister Albert Speer broadcast live to the German people the following address:

Never before has a cultured people been smitten as grievously as the German people now! Never before has any land been so laid waste by the fury of the war as has Germany!

After being thoroughly searched, Gen. Alfred Jodl, Speer, and Dönitz are taken to the police station yard to be presented to the Allied press. (Photo by AFPU Capt. Bill Malandine; LC)

You are all disheartened now, and incensed. Instead of faith, desperation has entered your hearts; you have become tired and cynical. This must not be! The bearing of the German nation in this war has been such that, in times to come, future generations will look upon it with admiration. Let us not stop to cry out our eyes about the past—to work!

The havoc wrought by this war has only one parallel in history— the 30 Years' War. Yet the decimation of the people by starvation and plagues must not be allowed to reach the proportions of that period.

Speer, Dönitz, and Jodl were put on show to Allied photographers and war correspondents immediately after their arrest on May 23, 1945, the same day that Himmler committed suicide. (LC)

The massive amount of paperwork generated by the trial is evidenced here, with secretaries of the International Military Tribunal typing pool seen at work. During their time at Nuremberg, the Nazis were allowed to write their memoirs if they wished, and four did: Keitel, von Ribbentrop, Rosenberg, and Frank. All four were convicted and hanged. (Ray D'Addario)

That—and that alone—is the reason why Admiral Dönitz has resolved not to lay down arms. This is the only meaning of the struggle: to prevent the death of fleeing German men.

It rests with our enemies to decide whether they wish to grant to the German people the possibilities that lie open to a nation which is defeated, but which has shown its heroic spirit in battle, and imprinted its reputation on the pages of history as a generous and decent opponent.

Yet each one of us must contribute his share, and in months to come devote our strength to the work of reconstruction. You must overcome your lethargy, your paralyzing despair.[1]

Germany surrendered unconditionally on May 7–8, 1945. According to writer Karel Margry, "Speer alone argued that it was essential to form a government acceptable to the Allies, but none of the others could believe that they had done anything wrong under Hitler, or that the Allies could possibly object to them."[2]

Astute politician Dr. Speer knew better. Because it had been a regime appointed by Hitler's hand-picked successor, he believed that the Acting Reich Government would be turned out of office sooner or later, and, indeed, it lasted a mere 21 days. No doubt Professor Speer saw this as his moment of destiny, the chance that he'd been waiting and planning for since at least the loss of Stalingrad in February 1943: to be the new, postwar leader of the Fourth Reich—under Allied supervision, of course.

According to his memoirs: "My tart suggestion that a few leading Social Democrats and liberals be brought forth to take over our functions went unnoticed."[3] Thus, just as Stalin had succeeded the dead Lenin, so, too, would Speer, in effect, have taken over the Reich from his late Führer, following the brief Dönitz interregnum. Meanwhile, according to Margry, Speer took as his official headquarters "Glücksburg Castle, four miles west of Flensburg, residence of the Duke of Mecklenburg-Holstein."

As Professor Speer himself recalled:

My adjutant reported that many American officers, including a high-ranking general, had arrived at the entrance to the castle. Our guard of soldiers from a German armored force presented arms, and so—under the protection of German arms, as it were—General F. L. Anderson, commander of the bombers of the American Eighth Air Force, entered my apartment. He thanked me in the most courteous fashion for taking part in the discussions.

The place where my family was staying was 25 miles from Glücksburg. Since the worst that could happen was that I would be arrested a few days earlier, I drove out of the enclave around Flensburg and—thanks to the careless unconcern of the British— reached the occupied zone without trouble, The British soldiers who were strolling in the streets paid no attention to my car.

Heavy tanks stood in the villages, their cannon protected by canvas hoods, so I arrived safely at the door of the country house where my family was staying. We were all delighted at this prank, which I was able to repeat several times, but perhaps I strained British nonchalance after all.

On May 21, I was taken back to Flensburg in my car, and locked in a room at Secret Service headquarters, watched over by a soldier with an automatic rifle on his knees. After a few hours, I was released. My car had vanished; the British took me back to Glücksburg in one of their cars.[4]

During the third week of the Dönitz government's existence, Speer was "interrogated" by two directors of the US Strategic Bombing Survey. Their article discussed how they felt Speer had already prepared his strategy, and his story, to ensure his survival:

Speer had a plan for establishing an "in" with the winning side—what amounted to a well-devised strategy of self-vindication and survival.

The first part of Speer's strategy was to qualify himself as a brilliant technician and administrator. He could guess that his enemies admired brains and technical ability. He could at least hope that he

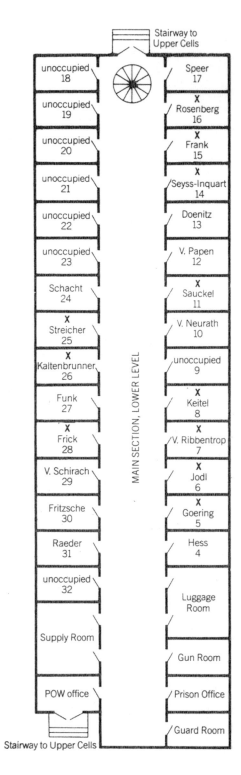

An excellent floor plan of the prisoners' wing at the Nuremberg Prison. Speer's cell, #17, is at the upper right. On August 8, 1952 at Spandau, Speer read Dr. Gilbert's *Nuremberg Diary*, commenting: "Reading about oneself as if one were long dead." (Library of Congress, Washington, DC)

A good interior view of the prisoners' wing ground floor and upper tiers, complete with US Army Military Policemen. (Ray D'Addario)

IMT defendant Karl Dönitz in his cell at the Nuremberg Palace of Justice during his trial as a war criminal. Convicted, he was sentenced to 10 years' imprisonment at Spandau outside Berlin. The feisty Nazi Grand Admiral lost two sons in U-boats during the war. (US Army Signal Corps photo by Ray D'Addario, Holyoke, MA)

had abilities they might want. The second part of his strategy was to appear completely unconcerned over his own fate. No one admires a coward; Speer wanted us to know that he realized his danger, and did not care. He joked about the way defeat had interrupted his career. Once, he asked a member of our party to arrange for his arrest—he said he was embarrassed by being in the Dönitz government, which he characterized as "poor opera." On another occasion, he jokingly invited one of the writers—who he had learned was a lawyer—to defend him at the war criminal trials. "Where," asked Speer, "could you get a more famous client?" Most of the Nazi bigwigs were in a

Stripped of his SS regalia and rank tabs, Dr. Brandt stands in the dock of the Doctors' Trial at Nuremberg's Palace of Justice. Convicted of war crimes, he was hanged. On December 9, 1946, Dr. Speer noted in his prison diary: "I sometimes see Karl Brandt among the defendants in the doctors' trial. One of the reasons that I flew back into burning Berlin in April 1945 was to save him from Hitler's death sentence. Today he waved sadly to me in passing. I hear there is gravely incriminating evidence of his having engaged in medical experiments on human beings. I frequently sat with Brandt; we talked about Hitler, made fun of Göring, and expressed indignation at the sybaritic living in Hitler's entourage, at the many parasites. But he never gave me any information about his doings, any more than I would have revealed to him that we were working on rockets that were supposed to reduce London to rubble. Even when we spoke of our own dead, we used the term 'casualties,' and in general, we were great at inventing euphemisms"— perhaps also like the Final Solution of the Jewish Question. (Photo by Ray D'Addario, US Army Signal Corps, Holyoke, MA)

state of complete moral collapse at the time of the surrender. Speer knew that he looked good by comparison...

The third part of Speer's strategy was to present himself as a great humanitarian. He claimed to have given up all hope of victory in the spring of 1944, and to have devoted himself thereafter to saving his people from unnecessary suffering. He had especially resisted efforts by the other leaders to adopt a scorched earth policy within Germany...

Closely related to Speer's effort to picture himself as a humanitarian was an effort to disassociate himself from his former colleagues. It was the vigor with which he attacked them that first caused us to suspect that Speer was rehearsing the story with which he hoped to save his neck. Not even the British or American press ever did better. Göring Speer described as an acquisitive looter and a dishonest, sensual, and stupid traitor; Goebbels he described as an empty-headed fool; his propaganda succeeded only in convincing the German leaders that they could win the war without exerting themselves. Bormann he characterized as a vicious gutter politician, Ribbentrop a clown, Himmler a monster, and Sauckel a beast... He especially assailed the moral corruption of his colleagues...[5]

He also provided damning condemnation of the German government in the last months of the war:

[Hitler's] bunker, Speer suggested, symbolized the total divorce of government from reality. "There were no windows or doors; even the air was pumped in." Field Marshal Keitel and General Jodl, he charged, knew nothing of conditions at the front.

The men in the bunker spent their time looking for scapegoats for the defeats, or indulging in the childish hope that some magic plan

Accused war criminal defendant Erich Raeder in his cell at Nuremberg Jail, 1945. Note the US Army mess kit drinking cup on the table. (Photo by Ray D'Addario, US Army Signal Corps)

would bring victory. Field commanders were being replaced constantly. Plan after plan was discussed and discarded.[6]

Finally, "Speer was clever enough to realize that his attack on his fellow Nazis raised a large question, namely, 'Why, Mr. Reich Minister, did you keep such frightful company?' ... Speer covered this loophole with repeated references to his friendship with Hitler. It was because of his love for Hitler that he took and held office."[7]

The Allies decided at last to end the final charade of the Third Reich on May 23, 1945, the same day that Himmler committed suicide while a prisoner of war held by the British Army. Noted Margry:

The [British] Cheshire's Anti-Tank Platoon under Lieutenant F. B. Walker made its way to Glücksburg Castle, deployed its six-pounder guns on the banks of the surrounding lake, and stormed across the drawbridge that formed the castle's one and only entrance.

Having been told that the castle was the seat of the Dönitz government, the platoon had expected to find many cabinet ministers there, but this was not the case. However, they did catch one Grade I person: Albert Speer, who was found in a small cloakroom shaving. The British sergeant at the door said, "Are you Albert Speer, sir?" At Speer's reply that he was, the sergeant stood at attention, and said: "Sir, you are my prisoner," and told him to pack his things.[8]

Defendant Albert Speer (right) enters the dock of the accused from the small elevator at the rear, watched by both his fellow defendants and US Army MPs wearing white web belts. In the front row of the dock from left to right are Göring (a blanket over his lower torso), Hess staring vacantly into space, von Ribbentrop, and Keitel (turning to look at Speer); back row: Dönitz (dark glasses to shield his eyes from the bright photographic lights) listening to Raeder, von Schirach and Sauckel. Speer angered Göring by breaking the united front Göring forged among the other defendants to defend the Hitler regime against its Allied accusers. (Previously unpublished photo by Ray D'Addario, US Army Signal Corps, Holyoke, MA)

Above and above right: Speer's courtroom nemesis at Nuremberg was defendant Hermann Göring, seen here (right) stepping down from the witness box during his own trial; and in the witness box. At left is US Army psychologist Dr. Gustave M. Gilbert, author of the 1947 *Nuremberg Diary*. Once in the courtroom, an indignant Göring confronted Speer in the dock directly, and the latter told him to "Go to hell!" back in his cell, Göring referred to "That stupid fool Speer" and how his breaking of their united front was "Such a rotten thing to save his lousy neck—to put it bluntly—to piss in front and crap behind a little longer!" (Ray D'Addario)

Here is Professor Speer's version of events on May 23, 1945:

My adjutant came rushing into my bedroom. The British had surrounded Glücksburg. A sergeant entered my room and announced that I was a prisoner. He unbuckled his belt with its pistol, laid it casually on my table, and left the room to give me the opportunity to pack my things.

Soon afterward, a truck brought me back to Flensburg. As we rode off, I could see that many antitank guns were trained on Glücksburg Castle. They still thought I might be capable of far more than I was. Shortly afterward, the Reich War Flag, which had been raised every day at the Naval School, was taken down by the British.

If anything proved that the Dönitz government—try though it might—was not a new beginning, it was the persistence of this flag… At the beginning of our days in Flensburg, Dönitz and I had agreed that the flag must remain. We could not pretend to represent anything new, I thought. Flensburg was only the last stage of the Third Reich, nothing more."9

Thus, Dr. Albert Speer, a third-generation, upper-class architect until he joined the plebeian Nazis, had ended his professional career hobnobbing with right-wing, arch-conservative royalty, a class hated in the main by the more socialist Nazis. Margry concludes his narrative thus:

The Grade 1 prisoners were taken to the main police station at #1 Norderhofenden in the center of Flensburg. Formerly the Flensburger Hof hotel, the building had been a police and Gestapo section since 1935.

On May 14, 1945, [British] Brig. Jack Churcher had requisitioned this building as headquarters for his 159th Infantry Brigade… That afternoon Dönitz, Jodl, and Speer, and the other Grade 1 prisoners, were taken to the 159th Brigade headquarters at the police station on Norderhofenden in the center of Flensburg.

There they were sat in a room surrounded by their suitcases. One by one, they were summoned to an adjoining room to be registered as prisoners, and taken behind a screened area in the corner where they were subjected to a full body search. No crevice was left unexamined.

…The humiliating body search and the rifling of their luggage severely upset the Germans. Depending on their dispositions, they emerged from the room angry, insulted, or depressed.[10]

Here is how Speer remembered this humiliating incident 35 years later: "When my turn came, I, too, was affronted by the embarrassing physical examination to which I was subjected. Probably it was a consequence of Himmler's suicide; he had kept a poison pill concealed in his gum."[11]

Dönitz, Jodl, and Speer, the three main prisoners, were now taken to the rear courtyard of the police station where Allied reporters and photographers were waiting to film, photograph, and question them: "The three Germans took it calmly and stoically. Speer made an effort to give the impression that the spectacle did not concern him"—a veteran and determined play actor like Hitler in even the worst of circumstances, ready for the next round![12] Following this, Speer recollected:

… we were squeezed into several trucks along with the others from the waiting room. Ahead of us and behind us—as I could see at curves in

Accused war criminal Dr. Alfred Rosenberg (second from left) in the dock at Nuremberg before the International Military Tribunal (IMT), being addressed by prison psychologist Dr. Gustave M. Gilbert (right), author of *Nuremberg Diary*. The others are, from left to right: Col. Gen. Alfred Jodl, two US Army military policemen, former Nazi Governor-General of Poland Hans Frank (dark glasses), former Reich Protector of Holland Dr. Artur Seyss-Inquart, Professor Speer and former Minister of the Interior Dr. Wilhelm Frick (back to camera). (US Army Signal Corps photo by Ray D'Addario, Holyoke, MA.)

Outside the cells in the prisoners' wing of the IMT at Nuremberg Prison, 1945–46. Each cell had its own US Army MP standing guard. (US Army Signal Corps photo by Ray D'Addario)

the road, we had an escort of 30–40 armored vehicles, a rare honor for me, accustomed as I was to driving around in my car alone and without protection.

At an airport, we were loaded into two two-motored cargo planes... On only two of the [later] journeys was the end completely clear: the one to Nuremberg, and the one to Spandau... [The destination was] Luxembourg. The plane landed; outside, a cordon of American soldiers was drawn up in two rows. Each of them had his automatic rifle trained on the narrow lane down which we would walk between them.

I had seen such a reception only in gangster films when the criminals are finally led off to justice. In open trucks, seated on crude wooden benches and guarded by soldiers again with their guns at the ready, we were taken through several villages where the people in the streets whistled and shouted at us, epithets we could not make out. The first stage of my imprisonment had begun.

We stopped at a large building, the Palace Hotel in Mondorf, and were led into the lobby. From outside, we had been able to see Göring and other former members of the leadership of the Third Reich pacing back and forth. The whole hierarchy was there: ministers, field marshals, reichsleiters, state secretaries, and generals...[13]

Two weeks later, Speer was greeted by an American lieutenant with a car. They drove to Paris, which he had last seen with Hitler on June 23, 1940. They arrived at the Trianon Palace Hotel at Versailles: "I knew the way there well; it was the place where I had stayed in 1937 when I

The two American prosecutors at the IMT during 1945–46. At left is Brig. Gen. Telford Taylor, assistant prosecutor, and at right US Supreme Court Justice and IMT chief prosecutor Robert H. Jackson, who focused his main attention on Göring, not Speer. Justice Jackson respected Speer, and the latter was aware that it was his ability to be liked that carried him through the trial. At Nuremberg, after his sentencing, Speer took the time to read his judges' opinions on their sentences: "A whole book," he noted. (Ray D'Addario)

was designing the German pavilion for the Paris World's Fair."[14] Now it was Gen. Dwight D. Eisenhower's headquarters, but Dr. Speer was next taken to Chesnay instead, where he lived for several weeks.

"I ended up in a small room on the third floor of the rear wing. Its appointments were Spartan: an Army cot and a chair. The window was laced over with barbed wire. An armed guard was posted at the door."[15]

The other prisoners "were almost exclusively leading technicians and scientists, agricultural and railroad specialists, among them former Minister Dorpmüller … [and] Professor Heinkel… A week after my arrival, my permanent guard was withdrawn, and I was allowed to walk about freely…

New prisoners arrived: various members of my ministry, among them Frank and Saur. We were also joined by technical officers of the American and British forces, who wanted to expand their knowledge of German conditions. My assistants and I agreed that we ought to place our experience in the technology of armaments at their disposal.

I could not contribute very much; Saur had by far the better knowledge of details… A few days later, a large bus drew up in the prison yard… Schacht and General Thomas, the former Chief of the Armaments Office…

When Eisenhower's headquarters were shifted to Frankfurt, a column of some 10 American military trucks appeared at our quarters. We prisoners were assigned our places in two open trucks with wooden benches… As we passed through Paris, at every traffic stop, a crowd assembled shouting insults and threats.[16]

They then passed through his former home towns of both Heidelberg and Mannheim, then on to Nauheim, to Kransberg Castle…

It was the gathering point for all kinds of specialists: almost the entire leadership of my ministry, most of my department heads, most of the leading men in munitions, tanks, automobile, ship, aircraft, and textile production, the important figures in chemistry, and such designers as Professor Porsche…

Wernher von Braun and his assistants joined us a few days later. He had received offers from the United States and England for himself and his staff, and we discussed these. The Russians, too, had contrived to use the kitchen staff at the heavily guarded Garmisch camp to smuggle an offer of a contract to him.

For the rest, we banished boredom by early morning sports, and a series of scientific lectures, and once Schacht recited poetry, giving astonishingly emotional renderings. A weekly cabaret was also conjured up. We watched the performances—the scenes repeatedly dealt with our own situation—and sometimes tears of laughter ran down our faces at the tumble we had taken.

One morning … one of my former assistants roused me from sleep: "I've just heard on the radio that you and Schacht are going to be tried at Nuremberg!" I tried to keep my composure, but the news hit me

Accused Nazi war criminal defendants in the dock of the International Military Tribunal, Nuremberg, 1945, surrounded by US Army white-helmeted Military Police. In the front row at left, top to bottom. Göring, Hess, von Ribbentrop, and Keitel. Behind them in the second row, top to bottom, Dönitz, Raeder, Baldur von Schirach, and Fritz Sauckel, the latter two also Dr. Speer's wartime colleagues. (US Army Signal Corps photo by Ray D'Addario)

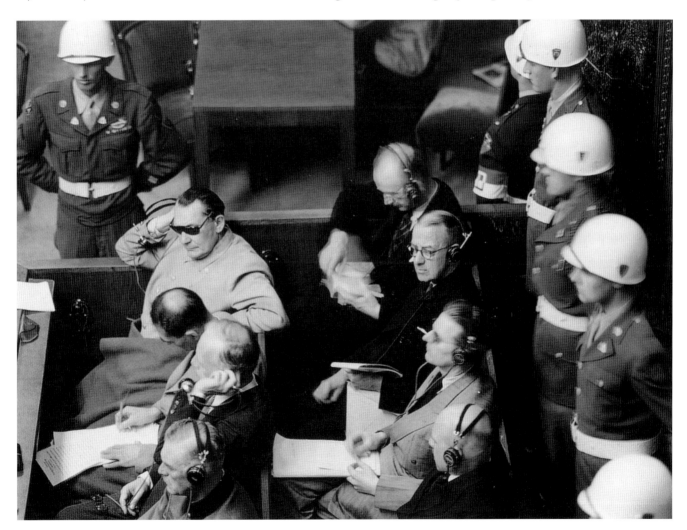

hard... Weeks were to pass before I was taken there... The Americans said cheerily, "You'll soon be acquitted, and the whole thing forgotten..."

An American jeep swung in through the gate: the squad that had come to get me. By the time I finally entered the jeep, almost the entire camp community had assembled in the castle yard. Everyone wished me well. I shall never forget the kindly and troubled expression in the eyes of the British colonel as he bade me goodbye.[17]

At length, Speer arrived at Nuremberg after dark one night. The German prisoner-of-war staffers were sympathetic, but there were no books or newspapers allowed. British Army Major Airey Neave—later an author and elected Member of Parliament—served the Allied indictment on Speer in his cell, and left an interesting account of his observations of the accused defendant in his 1978 work, *On Trial at Nuremberg*:

I remember the first sight of this gifted and compelling man.

He was waiting for me with a nervous smile. Behind that smile there was the patronizing look of the court-artist. I am glad we did not get to know each other well. His smoothness repelled me. It was he who made Nazism acceptable to the arts and sciences. Perhaps this was his greatest crime. Even today, he seeks to persuade us that he was "different," and, indeed, he was.

Defendant Karl Dönitz pleads not guilty, using a portable microphone in the second row, International Military Tribunal, Nuremberg, 1945. Front row, left to right: Göring, Hess, von Ribbentrop, Keitel, Ernst Kaltenbrunner, and Alfred Rosenberg. Second row, left to right: Dönitz (standing), Raeder, von Schirach, and Sauckel. (US Army Signal Corps photo by Ray D'Addario)

Speer was the strongest in character and most genuinely courageous of all the prisoners at Nuremberg, yet he was a man I could never trust… Speer was an impressive figure among the broken-down street politicians of the Nazi Party. His appearance was striking, even in his prison clothes. He was tall and dark with a strong, intelligent face. His manner was persuasive, and he seemed like an athletic university professor who had turned to public administration. His eyes were very large and thoughtful. He was, I felt, a man of considerable distinction…

I was at first impressed. Speer's charm and apparent integrity seemed to shine in that sordid place. He spoke to me in excellent English. He was at once frank about his responsibility. "The trial is necessary," he said as he took the indictment. "There is such a thing as common responsibility for such terrible crimes, even under an authoritarian system." He returned to this theme many times before the trial was over, both in and out of the dock, but I felt uneasy in his presence…[18]

This was, indeed, the entire linchpin of Speer's defense in a nutshell: to accept *overall* responsibility for the crimes of the regime, but to deny any *personal* culpability of his own. It was a very fine line, indeed, but defendant Speer managed to straddle that moral fence throughout the trial, and thus survived:

Speer has frankly admitted his mistakes and disillusionment. Compared with other Nuremberg defendants, he was candor itself… The greatest compliment was paid by Göring: "We should never have trusted him!" All this I understood directly I talked to him.

He seemed the one civilized man I had so far met that afternoon, and yet there was a darker side to this good-looking, friendly ex-minister… He was, I thought, more to blame than the miserable Sauckel for the sufferings of the foreign workers. There are those who regard him as "The real criminal of Nazi Germany" who for 10 years sat "at the very center of political power, and did nothing."[19]

What has come to be called the "Nuremberg Trials" is, in fact, a misnomer. There were 13 trials in all, but "The Trial of the Major Nazi War Criminals at Nuremberg" was the one concerned with the individual trials of the first 21 top leaders of the regime, with the missing—but dead—Martin Bormann being tried *in absentia* as the 22nd defendant.

This first trial came to be called the International Military Tribunal, and lasted from November 20, 1945 through October 1, 1946 in Room 600 of the Nuremberg Palace of Justice, a spacious building that had 22,000 square meters of space with 530 offices and 80 separate courtrooms overall. It had received little structural damage during the war, and had a large prison attached to the rear that was part of the entire complex. It was ideal.

There were four counts in the Allied indictment: 1) Conspiracy to commit crimes against peace; 2) Planning, initiating, and waging wars of aggression; 3) War crimes; and 4) Crimes against humanity. Speer was indicted on *all four* counts. Over the 218 trial days, testimony was introduced from 360 witnesses—in both verbal and written form, some 236 witnesses from the court itself, and some from judges assigned to take testimony. In addition, 200,000 affidavits were also evaluated as evidence under Anglo-Saxon legal procedures. More than a thousand personnel staffed the IMT as text translators, simultaneous translators, secretaries, etc.

No subsequent international tribunal took place. Between 1947 and 1949, 12 US military trials took place at Nuremberg, involving politicians, military personnel, businessmen, industrialists, medical doctors, lawyers, and members of the German Foreign Office. Other trials were held elsewhere. The official record of the International Military Tribunal was published in 42 volumes, the so-called *Blue Series.*

Of Speer's final statement from the dock on August 31, 1946, Neave opined:

> The evidence was overwhelming. Speer's sounded a chilling alarm to the world: "Hitler's dictatorship was the first of an industrial state in this age of modern technology, a dictatorship which employed to perfection the instruments of technology to dominate its own people.

Defendant Speer, with a US Army MP standing behind him, talks with his defense counsel Dr. Hans Flächsner (right), and that of Dönitz, Otto Kranzbühler. By disregarding their expert legal advice, Speer risked being hanged by the Allies, and nearly was, too. Speer recalled: "My discussions with Flächsner in the Nuremberg courthouse took place with the two of us separated by a fine-meshed wire grating and watched constantly by an American soldier. At the bottom left of the wooden frame was a slot through which the lawyer could exchange documents with his client; but all papers had to be received by the soldier first." (Ray D'Addario)

"By such instruments as the radio and public address systems, 80 million people could be made subject to the will of one individual. Telephone, teletype, and radio made it possible to transmit commands from the highest to the lowest levels, where they were executed uncritically.

"Thus, many officers and squads received their evil commands in this direct manner. The instruments of technology made it possible to keep a close watch over all citizens, and to keep criminal operations shrouded in a high degree of secrecy. The outside of this State apparatus may look like the seemingly wild tangle of cables in a telephone exchange, but like such an exchange, it can be directed by a single will."[20]

Other authors and eyewitnesses also left their impressions of Speer, and a selection of their findings is worth noting. United States Army Col. Burton C. Andrus was the prison commandant, and wrote a volume of memoirs entitled *I Was the Nuremberg Jailer*.[21] When the famous "atrocity film" was shown in the courtroom, Colonel Andrus noted the defendants' various reactions: "Speer, looking sad, swallowed often." Later:

> Speer was grumbling openly about Göring's nettling accusations that Speer had "sold out" to the prosecution, and that [Dr. Gustave M.] Gilbert had influenced the weaklings.
>
> Speer sought Gilbert's commiseration: "Can you imagine!" he said. They can't conceive of anybody telling the truth. Everything is crookedness and "deals" because that is the only way they can think after they acted that way all their lives.[22]

Speer's attorney—Dr. Hans Flächsner—assessed that his client was "manipulative ... a game player who used people, didn't love anyone: he was still what he had always been."[23] I concur.

Psychiatrist Dr. Douglas M. Kelley found that Speer, "Was a frank, friendly personality. He is oddly enough more interested in architecture than in the trial. He has an excellent outlet for his emotional stress in his drawing. He is calm, resigned, well adjusted, and without tension."[24] Speer wrote on July 11, 1947: "Have just read what Kelley has to say about me. In his opinion, I am one of the most servile, but highly intelligent; also, a very talented architect who devoted himself solely to his work with boyish enthusiasm—a racehorse with blinkers. Now in prison, I am inhibited, he says, but still candid."[25]

John K. Lattimer, MD was also at Nuremberg. In his book *Hitler's Fatal Sickness and Other Secrets of the Nazi Leaders*, Dr. Lattimer wrote:

> Albert Speer—the quiet, reserved, shaggy-browed architect—was 40 years old when the trial started... He was certainly the most attractive of the group, and proved to be a very capable person ... was well educated and very intelligent in a pragmatic way.

	Charges counts				Verdicts counts				Sentences
	1	2	3	4	1	2	3	4	
Hermann Göring	X	X	X	X	X	X	X	X	Death
Joachim von Ribbentrop	X	X	X	X	X	X	X	X	Death
Wilhelm Keitel	X	X	X	X	X	X	X	X	Death
Alfred Jodl	X	X	X	X	X	X	X	X	Death
Alfred Rosenberg	X	X	X	X	X	X	X	X	Death
Wilhelm Frick	X	X	X			X	X	X	Death
Artur Seyss-Inquart	X	X	X			X	X	X	Death
Fritz Sauckel	X	X	X				X	X	Death
Martin Bormann	X		X	X			X	X	Death
Ernst Kaltenbrunner	X		X	X			X	X	Death
Hans Frank	X		X	X			X	X	Death
Julius Streicher	X			X				X	Death
Erich Raeder	X	X	X	X	X	X	X		Life imprisonment
Walther Funk	X	X	X			X	X	X	Life imprisonment
Rudolf Hess	X	X	X	X	X	X			Life imprisonment
Albert Speer	X	X	X				X	X	20 years
Baldur von Schirach	X			X				X	20 years
Constantin von Neurath	X	X	X	X	X	X	X	X	15 years
Karl Dönitz	X	X	X			X	X		10 years
Hans Fritzsche	X		X	X					Acquitted
Franz von Papen	X	X							Acquitted
Hjalmar Schacht	X			X					Acquitted

A chart showing the defendants, charges, verdicts, and sentences. Speer recalled that he never forgot how he heard his sentence: "'Albert Speer, to 20 years' imprisonment,' like yesterday," handcuffed to a US Army MP. (*After the Battle* magazine)

His manner was quiet, boyish, almost shy, and he got along with everybody, complying with all of the onerous regulations without difficulty, and with unfailing politeness. Speer was fluent in several languages, and often assisted the court interpreters when they stumbled over a word while trying to keep up with the difficult technical testimony. Leaning forward in his seat, he would look at the interpreters and shake his head 'no,' indicating an error, which they would then hasten to correct. If they still had trouble, he would inconspicuously pass them a note with the correction...

In the end, Speer stoically accepted his sentence, looking the judge right in the eye... It seemed to me that he made the mistake of talking too much, rather than just answering questions, as all good lawyers advise. It occurred to many of us that his talents should have been put to work to alleviate the extreme hardship suffered by the German populace as a result of our devastating bombing.

This, his own tendency to self-punishment, denied his people his services in the reconstruction period after the war. It was almost as if he demanded to be punished... As it happened, Speer knew a good deal about Charles Lindbergh, and admired him greatly. We discussed him at length. They were both technological wizards, but very quiet, very pleasant people...

In the final analysis, Speer turned out to be by far the most likable and capable of the prisoners. He was always helpful, always amiable, and had shown tremendous capability as a manager of German war production.[26]

As noted by Assistant Trial Counsel Drexel A. Sprecher of the International Military Tribunal at Nuremberg in a letter to the editor of *The Washington Post* on September 8, 1981:

> Mr. Speer surprised those of us who were prosecutors by his consistency in admitting a considerable degree of responsibility and guilt for his role as one of the principal leaders of the Nazi regime during World War II...
>
> In his final statement, he said: "A new large-scale war will end with the destruction of human culture and civilization. Nothing can prevent unconfined engineering and science from completing the work of destroying human beings, which it has begun in so dreadful a way in this war.
>
> Therefore, this trial must contribute toward preventing such degenerate wars in the future, and toward establishing rules whereby human beings can live together. Of what importance is my own fate—after everything that has happened—in comparison with this high goal?"[27]

Assistant US Prosecutor Brig. Gen. Telford Taylor assessed Dr. Speer in his 1992 work *The Anatomy of the Nuremberg Trials*: "There was a general belief that the evidence against him was heavy—as heavy, for example, as Sauckel's. Speer could not help himself by arguing his innocence: his cards were his defiance of Hitler's scorched earth orders and his own personality." As for defendant Speer's celebrated final statement, Brigadier General Taylor noted: "Rereading it today, his words to me seem less impressive, and Speer's grand indifference to his own fate too studied."[28]

Brigadier General Taylor concluded that one reason Speer wasn't hanged was because—in 1945—he had turned away from his loyalty to Hitler. He also noted that the French and British judges had favored Speer, but that the American and Russian judges both wanted him hanged. Then the American judge switched to a 20-year sentence instead of 15 years, and thus Speer escaped the hangman's noose.

Writing in his 1977 study *Reaching Judgment at Nuremberg*, author Bradley F. Smith termed Speer's a "difficult" verdict: "The adaptable Funk and Speer went to jail, while [others] went to the gallows."[29] The court's "social prejudices" helped Speer to survive, as several of the Western members actually liked him, as he had, indeed, calculated.

There was also Speer's alleged plot to gas the Führer and his entourage in 1945 in the Berlin underground shelter that he himself had installed, starting in 1943. According to Smith:

> Speer also informed the Court—haltingly, and perhaps coyly—that in the last days he had toyed with farfetched, if not harebrained plans to kill Hitler, but Speer's very ability to see and act in this manner also cut

against the force of mitigation. From the very beginning, Speer has possessed the position—and the education—to grasp the moral implications of what he was doing.

The key to the Speer judgment was whether the Court would hold that his professional status and education did more to incriminate him during the first 95% of his career than to exonerate him in the last 5%.[30]

While his fellow Nazis in the dock were infuriated that he had even *considered* murdering their Führer, the Western Allies *bought* this defensive tactic, while the Soviets never did. They saw Speer for what he was: a liar. As for me, I believe that this "Bunker Plot" was entirely bogus, invented by Speer—and known only to him—as a clever stratagem that worked.

Before that, however, Speer had already secretly made an overture to Justice Robert H. Jackson in the form of a personal letter to him of November 17, 1945, just as the IMT trial was beginning. In it he stated that he had certain information, "As to military, technical questions that should not be made known to other persons," such as the Soviets, the emergent Western enemies of the Cold War that Hitler and Göring had always predicted would come in the wake of the end of World War II. Did Justice Jackson take the bait?[31] Maybe, but the salient point is that defendant Speer was politically shrewd enough to make this inspired gambit, at the very outset of the Cold War that would dominate the next 50 years of East–West relations, and this, too, was a cornerstone of his successful defense.

There was another side to Speer's Jackson gambit, as discussed in author Joseph E. Persico's excellent study, *Nuremberg: Infamy on Trial*:

> Tom Dodd had been preparing himself for the Albert Speer cross-examination, when Speer asked if he might see the deputy prosecutor in the visitors' room. Speer told Dodd that Hermann Göring was his chief rival for the soul of the defendants. Göring stood for truculent defiance. Speer stood for admission of Nazi guilt. Göring had been cross-examined by Jackson, the chief Nuremberg prosecutor, but Speer was going to be questioned by a subordinate.
>
> With all due respect to Dodd, would not this difference be noticed by the other defendants? And would it not put Speer—in their eyes—in an inferior status to Göring, thus making it more difficult for Speer to win them to his side?
>
> Dodd was perplexed by Speer's peculiar measure of status. He nevertheless took up the matter with Jackson, and recommended that the chief prosecutor take over Speer's cross-examination. If it made the man feel more important, if it made him a more cooperative and useful witness, why not? He had no ego at stake in this assignment, Dodd said. Jackson agreed.
>
> When word of the change got out, some of the prosecution staff became suspicious. Speer sat in court every day. He could not be unaware

US Army hangman at Nuremberg, Sgt John Woods, who died later in the Korean War. He was quoted as saying, "I'm glad that I hanged those Nazis." Göring cheated his hangman by taking cyanide in his maximum-security cell shortly before the scheduled hanging on October 16, 1946. (US Army Signal Corps)

that Dodd was a tough, skilled, dangerous prosecutor. Jackson's performance in the adversarial arena was of another caliber, and Speer knew it.[32]

Speer also knew—as did Jackson himself—of the secret letter between them, a knowledge none of the others in the courtroom on either side shared. The deal was in place. Jackson would go lightly on Speer, and Speer would live.

For their part, the Soviets wanted Speer convicted on all four counts—and hanged—but the Western Allies spared his life, keeping him in reserve as a bargaining chip and, indeed, even a foil, for possible future usage. Thus was Speer's life saved in this behind-the-scenes pretrial deal. On September 11, 1946, therefore, the Western Allied justices voted for the controversial 20-year sentence, not the noose of the gallows.

The two men who from the first realized the full stakes over which the trial would be fought were the chief antagonists of the opposing points of view: Göring and Speer. Göring took the stance of defending Hitler and the Third Reich against the hated foreigners and the alleged "victors' justice"—thereby condemning himself to death irrevocably— while Speer opted for life instead. In a sense, both men won.

In her superb work *Eyewitnesses at Nuremberg*, English author Hilary Gaskin wrote: "Göring, Ribbentrop, and Speer were probably the most alert."[33] Once he had perceived the true Cold War inner dynamics of the IMT, Speer was polite to the Western Allies and virtually contemptuous of their—and his—rivals, the Soviets, even once crossing swords in the courtroom with Russian prosecutor M. Y. Raginsky, as noted by Eugene Davidson:

> Raginsky asked him if it was not true that he had given himself without reservation to his war tasks. Speer: "Yes, I believe that was the custom in *your* State, too." Raginsky: "I am not asking you about our State! We are now talking about your State!" Speer: "Yes. I only wanted to explain this to you, because apparently you do not appreciate why, in time of war, one should accept the post of Armaments Minister. If the need arises that is a matter of course, and I cannot understand why you do not appreciate that, and why you want to reproach me for it." Raginsky: "I understand you perfectly!" Speer: "*Good!*"[34]

As summarized by one source, "The verdicts were announced on September 30 and October 1, 1946: three acquittals, 12 death sentences by hanging, seven to life imprisonment, and others to lesser terms of 10 and 20 years, etc. Of the various Nazi organizations, guilty verdicts were handed down on the Nazi Party Leadership Corps, the SS, SD, and Gestapo," but, oddly, *not* on the SA Storm Troops that had helped hoist Hitler into office.[35] Speer was found guilty on Counts 3 and 4 of the Allied indictment, and was sentenced to 20 years' imprisonment.

CHAPTER ELEVEN

PRISONER #5: FROM "SPANDAUER" TO BESTSELLING AUTHOR, 1947–81

The fact that he served his entire sentence ... [is] a very rare case of justice served.
Norman J. W. Goda, *Tales from Spandau: Nazi Criminals and the Cold War*

Speer's release from Spandau after 20 years' imprisonment, 1966. (LC)

The reason that Speer *did* serve the total 20-year sentence was explained by author Norman Goda: "As luck would have it, this was due *not* to the Allies—who accepted Speer's Nuremberg story—but to the Soviets and the East Germans, who *never* did." *They* had wanted Speer hanged.

The East Germans even later feared that a released Speer—since he was still a young man—might become West Germany's Arms Minister under Chancellor Konrad Adenauer, and would build missile bases in the Free World that would have the same former Nazi aim of extending Western Europe to the Ural Mountains.

His fellow co-defendants also never accepted Speer's Nuremberg fairy tale, for they *knew* the true story, having seen it all unfold first-hand, especially Göring. At Spandau, Prisoner #5—as he was called—was generally shunned as a pariah by all six of the others, despite what he wrote of their relations generally in *Spandau: The Secret Diaries*.

The "Reich penitent's" basic stance at Spandau was that his conviction was "proper" (and he had miraculously been spared the gallows that he so deserved), and thus there was never any move for clemency or a pardon, but from day one there *was* an unceasing effort for his early release. Indeed, Speer instructed his daughter Hilde to forget about freeing Hess and von Schirach: "We fight only for me," as they no doubt suspected.

The Speer release team left no stone unturned, approaching the churches, the International Red Cross (including the Soviet version, the

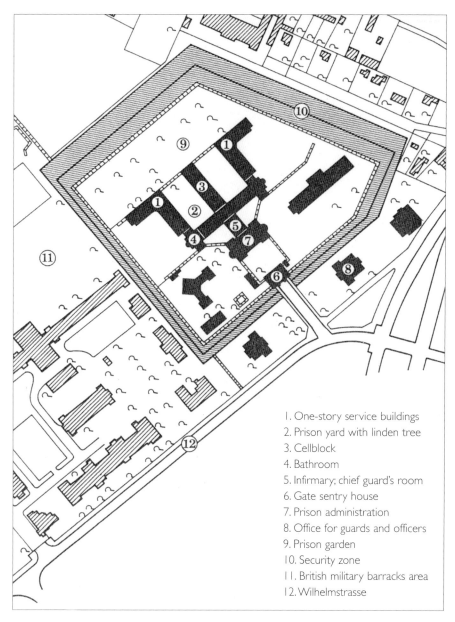

A diagram from Speer's own *Spandau: The Secret Diaries* showing the prison layout. The outer walls were protected by a high-voltage electric circuit which would be instantly fatal. The garden, around 7,200 square yards, was overgrown when Speer arrived. Speer set himself the task of transforming the plot into a little park in which he regularly exercised. Prisoner #5 added: "In beautiful sunshine and with a fresh breeze from the Wannsee, I took a Sunday walk of several hours… By steadily increasing the distance I do each day, I have made up for those weeks of confinement to my bed during the winter. My record for one day so far is 24.9km, my best pace 5.8km per hour." (*Spandau: the secret diaries*)

1. One-story service buildings
2. Prison yard with linden tree
3. Cellblock
4. Bathroom
5. Infirmary; chief guard's room
6. Gate sentry house
7. Prison administration
8. Office for guards and officers
9. Prison garden
10. Security zone
11. British military barracks area
12. Wilhelmstrasse

Red Crescent), Chancellor Adenauer, French President General Charles de Gaulle, and, in 1964, US Attorney General Robert F. Kennedy.

The deal that the Soviets were willing to consider—according to Professor Norman J. W. Goda—was to close Spandau altogether in return for an Allied withdrawal from West Berlin, but President Dwight D. Eisenhower balked at this. As for Chancellor Adenauer, his heart "was never in the Speer case."

In his diary entry for July 19, 1947, Speer wrote:

Yesterday—Friday, July 18th—we were awakened at 4a.m. A platoon of American soldiers stood around in the prison corridor… Each of us handcuffed to a soldier, we left the prison, entered two ambulances, and drove out through the prison gate, accompanied by a convoy of personnel carriers…

I was given a window seat in a fast, comfortable passenger plane, my guard beside me. After my long imprisonment, this flight in glorious weather was a stirring experience... Because life had stood still around me during the recent past, I had lost awareness of the fact that it was going on outside. A moving train, a tugboat on the Elbe, a smoking factory chimney gave me little thrills...

At half-past eight, the handcuffs were snapped on with a quiet click. From the descending plane, I could see a column of cars and many soldiers start into motion. We entered a bus whose windows were painted black... We got out. Behind us, at the same moment, the gate closed in a medieval-looking entrance... A command rang out in English: "Take off those handcuffs. None of that here." With a certain solemnity, the American guard shook hands with me in parting...

We were wearing our own clothing... Now, one by one, we filed into a room where we exchanged these clothes for long, blue convict's trousers, a tattered convict's jacket, a coarse shirt, and a convict's cap. The shoes are canvas, with a thick wooden sole. We are being given the clothing of convicts from concentration camps. The order of our entry decided our prison numbers. From now on, therefore, I am #5... My cell is 3 meters long and 2.7 meters wide... It is also 4 meters high...

A smiling Minister Speer driving an SdKfz 2 Kettenkrad (tracked motor cycle) which was used as a light tractor, through deep mud at a test site in Thuringia during the war. (HHA)

As in Nuremberg, the glass of the window has been replaced by cloudy, brownish celluloid… But when I stand on my plain wooden stool and open the transom of the window, I see the top of an old acacia through the stout iron bars, and at night the stars.[1]

Spandau prison was originally designed to imprison 600 people, after Nuremberg, it held just seven. *Berlin Then and Now,* says of it "This 19th century prison building would have left no mark on history if it had not been chosen for the imprisonment of the seven Nazi war criminals given sentences by the International Military Tribunal at Nuremberg in 1946. It was conveniently surrounded by military barracks and lay in the British Sector of West Berlin. However, from the time the prisoners arrived on July 18, 1947, it would be closely watched by all four wartime Allies over the next 40 years."[2] Fishman listed the Spandau Prison staff in 1954 as 10 waiters, 11 cooks, 14 kitchen helpers, three housekeepers, 14 charwomen, five clerical staff, and 29 administrators. The cost of running the site he calculated at $62,200 in 1954 dollars.[3]

Speer began reading and writing by "Transforming a prison cell into a scholar's den." Over the months and years ahead, he noted also,

Hitler's 52nd birthday, celebrated at FHQ Wolf's Lair, Rastenburg, East Prussia, April 20, 1942. From left to right: Chief of the German Army General Staff Col. Gen. Franz Halder, Keitel, Hitler, Reich Marshal Hermann Göring, Dr. Speer (obscured), and Grand Admiral Dr. Raeder. (Previously unpublished photo from the Hermann Göring Albums, Library of Congress, Washington, DC)

their guards went from hating them to treating them as fellow humans in a difficult situation.

Speer noted on November 2, 1949: "Family photos accompany every letter from Heidelberg. I hold them in my hand again and again, comparing them with older pictures—at least in this way, I can follow the children's development. I have been puzzling over one of the photos that came today: the forehead is obviously Fritz's and the haircut also suggests one of the boys, but the chin looks like Hilde's and the eyes are altogether Margaret's. Could it be Margaret after all, with her hair cut short? I hope it doesn't turn out to be Ernst. Only recently, I mistook Ernst for Arnold. More than four years have gone by since I last saw them. Ernst [the youngest] was 1½ then; today, he is over five. Arnold is entering his teens, and Albert is already 15! In another five years, three of my children will be adults."[4]

A staff of 32 people looked after the seven prisoners. He noted that, "The lives of others take the place of our own," such as the birth of a guard's daughter. "I might be around for her marriage… We become involved in the lives of our guards," he wrote on February 12, 1948.[5]

"What has become the most important to me is the world of books," he wrote, being allowed to read each night from 6–10pm, when it was time for lights out, but only allowed to read about things prior to the First World War by prison regulations.[6] Forbidden to know about current events, the prisoners were informed by their guards anyway. "The bathroom is the principal depot for news," he wrote on October 30, 1950, during the Korean War.[7]

Time hung heavy on his hands. When they planted a chestnut sapling tree, prisoner Walther Funk said to Speer, "We'll be here to sit in the shadow of this tree."[8] On March 8, 1953, he wrote in his diary, "I began writing down my memoirs."[9] The following June 6, he transferred his parents' estate to his children for the funding of their college educations.

Inside the prison, their letters sent and received were both read and censored by the prison administrators. Professor Goda noted: "Both Wolters and Kempf … served as clearinghouses for Speer's massive secret correspondence smuggled out of Spandau by the prison orderly Toni Proost, and later by fellow Dutch orderly Jan Boon."[10]

Rudolf Wolters had been the keeper of Minister Speer's office chronicle during the latter's years as an appointed ministry head. He also set up the so-called Schulgeldkondo (School Fund) in summer 1948 that sustained Speer's large family during the postwar years from donations made by his patron's former workers, associates, and German industrialists. Indeed, Prisoner #5 sent Wolters an actual *list* of 12 major figures owing him wartime favors from whom to solicit funds, with Rohland acting as the actual solicitor. When Frau Speer needed money for family expenses—and at her prisoner husband's discretion from jail—she simply made withdrawals. The family lived at the Speer parents' family estate at Heidelberg. During the 20 years that he served as both Speer's agent and literary conduit, Wolters marveled at "Father's political acumen" that he'd *always* had. Having lost faith in his IMT attorney Flächsner, Prisoner #5 instructed Wolters to assemble a new legal team to help effect his early release. He did so, but all efforts failed, as well as those of former Speer secretary Annemarie Kempf and Hilde Speer Schramm, the penitent's eldest daughter. A final School Fund withdrawal of 25,000 DM was raised by Wolters and Rohland as a cash gift upon Speer's release from Spandau (he promptly bought a sportscar, and gave his wife a watch!). In all, under Wolters' able direction, the School Fund accrued 150,000 DM. Speer failed to acknowledge any of Wolters' support in his books, and also betrayed what Wolters felt was their joint faith in Hitler as the leader of Germany before and during the war. Their final, "bitter" break occurred in 1971. Of this last, author Goda attests that Wolters "had grown weary of the discrepancy between Speer's comfortable lifestyle and the 'sackcloth and ashes' story that had made it possible."

"Wolters' secretary Marion Reisser typed thousands of pages from 25,000 scraps of Speer's scrawl smuggled out of Spandau, including letters to family, orders, and strategies for attempting his release, management of assets, and material eventually used in his memoirs and to create his *Spandau Diaries*."[11] I think it is significant that he uses the word "create" to describe the "diaries," rather than the word "kept," which is normally employed. Colonel Eugene Bird—a former US Army commandant of Spandau Prison in the 1960s—later recalled: "We suspected that he was writing his memoirs, when one day a rubber band around his trouser leg broke," revealing some hidden pages.

In 1953, a series of newspaper articles by Jack Fishman entitled *The Seven Men of Spandau* reminded the outside world that they were still there, and the following year his book of that same title also appeared. Also in 1954, a picture of the seven men published outside caused an international sensation. That same year, they were allowed to read daily newspapers to keep up with the new West Germany, and Speer daily devoured four. This obviated any desire by him to write a book after his release on how Germany's changes during his 20-year incarceration had affected him. Had he stepped outside those gates on October 1, 1966 "cold," there might well have been such a volume, I believe.

On November 12, 1954, Prisoner #5 was diagnosed with a pulmonary infarction, and he was briefly hospitalized. By the 29th, he had finished what came to be called the "Spandau draft" of his later memoirs, and had already decided to write a second on his prison years as well.

On March 19, 1955, he turned 50. On December 4, 1956, he noted that "My only subject is my past," and that fact remained true until the very day he died. On August 12, 1956, the prisoners were issued mosquito netting for their windows—after 10 years.

The Führer and his entourage ride in a FAMO SdKfz 9 heavy prime mover halftrack. Speer at right, Hitler at center, and Keitel at far left. (Previously unpublished photograph, HHA)

On July 14, 1957, he noted that, due to prison roof repairs: "For the first time in 10 years, I am experiencing evenings in the open air… How I would like—just once—to go walking in the moonlight."[12] The start of the East–West Berlin Crisis placed Spandau at the very vortex of world events, and Speer feared that they might all be taken to Russia on the Soviet watch as a result, but that never happened.

In 1959, he told Hess that he was keeping a diary. That June, prisoners and guards were forbidden to sit on the same bench together, as a warning against too much fraternization. In 1960, he noted that, "The idea of spending the rest of my days here is no longer frightening… There is great peacefulness in the thought. It is a matter of not seizing fate by the throat, but of willingly putting oneself in its hands."[13]

On August 21, 1960, he stated that they were to be permitted to listen to three hours of recorded music a month, but the very next day the Adolf Eichmann capture dashed their hopes of early release as the world focused anew on the barbarity of Nazi wartime crimes. On July 5, 1961, Speer wrote of his time in the garden, "We have a hedgehog. It lets me get close."[14] Later, he lamented, "Stricter regulations—after 15 years!"[15]

On November 12, 1961, he stated: "For the first time, I was alone with my wife … after more than 16 years… I could have held her hand,

could have embraced her. I did not do so. Sadot risked more than his job…"[16] On December 18, he shook hands with his son and namesake, Albert Jr.—and was caught doing so.

For New Year's 1962, a guard brought him a lobster and a bottle of British ale. Later, a guard gave him a Japanese-made Sony transistor radio: "For the first time in 17 years, I participated in a musical event," by listening to a concert.[17] Of the Cuban Missile Crisis, Speer noted on October 16, 1962, "Kennedy has ordered the total blockade of Cuba."[18] Two days later, he added, "This Cuban crisis is threatening the very life of the world."[19] When it was resolved, he asserted, "Peace! Very good."[20]

On May 16, 1963, he wrote of his pipe-smoking: "Today, after 18 years I stopped" to get in shape for his coming release— then still over three years away.[21] That October 22, he asked plaintively, "Will the third phase of my life be that of a scribbler?" as, indeed, it turned out to be.[22] That Christmas, he received his gifts from home in their wrappings for the first time, after 18 years.

On October 15, 1964, he wrote, "Khrushchev has fallen."[23] The West German Auschwitz trials of 1965 elicited this comment on January 12: "Suddenly, Spandau seems to me not so much the place of my imprisonment as of my protection."[24] That November 12, he noted that: "Indochina is now called Vietnam."[25] On December 5, 1965, he told his daughter Hilde that he "Will confront the press right away" at his release—as he did.[26]

Of his release he wrote:

A tower clock struck many times; then there was some bustle outside the cell. I was brought the old ski jacket in which I had been delivered

October 1, 1966: "Schirach shook hands with me in parting; we wished each other all the best for the future. A black Mercedes was waiting. Out of habit, I went to take my place in the front seat beside the driver, as I had always done in the past. Flächsner pushed me to the rear door; after all, I had to leave the prison at my wife's side." They went to Berlin's Hotel Gerhaus, where Speer gave a short press conference for the media, then retired for the evening. "I am quite glad to be out," he told the media. This was followed by two weeks of "quiet days" with his family in Schleswig-Holstein. According to Eric Norden in the *Playboy* interview, Speer "spoke briefly to the press, first in German and then in fluent French and English: 'My sentence was just. We were treated correctly and properly the whole time. I have no complaints.'" The Mercedes belonged to German steel executive and former Speer subordinate Ernst Wolf Mommsen. The day that both Dr. Speer and Baldur von Schirach were released, the Federal Republic of Germany decreed that neither would be paid a state pension. (LC)

to Spandau 19 years before, then an old tie, a shirt, and the corduroy trousers. I was taken to the directors. It was half-past eleven... The only formality was completed: I received 2,778 Reichs Marks, no longer valid, which had been taken from me in May 1945. I had no other possessions... At the stroke of 12, both wings of the gate were opened. Suddenly, we were bathed in blinding light. Many television spotlights were directed at us. In front—phantoms in the glare—I saw British soldiers running around us. For a moment, I thought I recognized Pease in the tumult, and waved to him. We passed through a thunderstorm of flashbulbs. The prison lay behind us. I did not dare look back.[27]

The first of his superb books—*Inside the Third Reich: Memoirs* (published in the United States in 1970)—was written at least in part in prison, then revised into book form upon Speer's release. It is an unrivaled, close-up view of the top stratum of the Nazi leadership corps in victory and defeat. The second tome—*Spandau: The Secret Diaries* (1976)—was essentially more of the same, interspersed with self-debates over the moral questions posed by the fate of the Jews and his own sellout to Hitler for a top spot among the chosen. The last book—*Infiltration: How Heinrich Himmler Schemed to Build an SS Industrial Empire* (also published under the title *The Slave State*)—was published the July before his sudden death, and was a detailed, if plodding, account of how the SS successfully invaded his production turf over the years.

Sprecher said of Speer's books in 1981: "Mr. Speer's best-seller books have added immeasurably to our knowledge of the psychology of persons who surround and support a modern dictatorship, and they have reduced the self-righteousness that afflicted so many early commentators on the Nazi regime!"[28]

Thus Dr. Speer entered the history books, initially mainly as seen in his own published works, plus the excellent *Playboy* magazine interview by Eric Norden in 1971, just before the release of his first book in English. The portrait that emerged was, by and large, that which Speer himself projected, i.e. as the Führer's favorite architect and minister, who—by sheer brilliance and superhuman effort—more than doubled German wartime armaments production. As the war was lost, he separated his loyalty from Hitler, and transferred it instead to Germany's survival after the war, and tried to bring the fighting in 1945 to as speedy a conclusion (with the minimum of internal destruction) as possible.

Later came the biographies, which presented different stories to that told by Speer himself, in particular that of Dr. Matthias Schmidt— *Albert Speer: The End of a Myth*—which Speer was attempting to block from publication at the time of his death.[29]

The book presents a well-considered challenge of some depth to Speer's own view of his desired—and rehabilitated—place in history, especially in the postwar era. The volume was hailed by many historians of the period as a powerful debunking of the "gospel according to Albert." It was Dr. Schmidt's central thesis that—far from being an "apolitical technocrat"—Dr. Speer participated up to the hilt in internal Nazi power politics, besting such ruthless luminaries as Göring, Goebbels, and Hess; strove to succeed Hitler as Führer; and worked in tandem with Himmler to build and maintain the extermination camps that promulgated the Nazi "Final Solution of the Jewish Problem."

Speer's indictment instead as a war criminal reportedly greatly stunned him, as he admitted, but—as Schmidt dryly noted—he adjusted, and set out to survive. Schmidt highlighted the fact that Speer's secret preparation of his first two books at Spandau was done with the aid of an associate whose name never appears in any of Speer's writings: Dr. Rudolf Wolters. He knew Speer from their student days in Berlin, and he kept the originals of the formal "Speer Office Journal" during the war. Schmidt produced photocopies of both the uncensored and later Speer-expurgated versions of these pages to show how the Nazi-turned-memoirist falsified his own history, which Schmidt called, "The most cunning apologia by any leading figure of the Third Reich."[30]

Speer died of a cerebral hemorrhage, in the arms of a woman other than his wife, on September 1, 1981—the 42nd anniversary of the German invasion of Poland—as the world pondered another possible violation of that unhappy country, and in a city, London, that V-1 flying bombs and V-2 rockets produced by his industrial combine had once tried to reduce to rubble. Dr. Albert Speer, 76—Hitler's former architect and Minister of Armaments and War Production, convicted war criminal, inmate of Berlin's Spandau Prison for 20 years, and best-selling author—no doubt would have appreciated the irony inherent in his own demise, for his writings are full of a subtle sense of irony and the cruel jokes that fate often plays on we mortals.

He would have appreciated, too, that—in death as well as in life—he would remain a controversial figure: damned by many, understood by some, and acknowledged by most historians as the pre-eminent memoirist of his era in history. Even the obituaries published the day after his sudden, unexpected passing unwittingly reported as fact some mistakes about his life.

For example, *The Washington Post* stated: "A courtly, patrician figure, he never joined the Nazi Party..."[31] *The New York Times* noted, "Mr. Speer was the only Nazi leader at Nuremberg ... in 1945–46 to admit his guilt."[32] This was not so; Hans Frank, Hitler's prewar personal lawyer and wartime Governor-General of Occupied Poland said: "A thousand years shall pass, and still the guilt of Germany shall not be erased." Frank, however, was a rather pedestrian sort, with the blood of several thousand slain Poles on his hands, almost predestined to hang by the victorious Allies at the war's end. Speer was, however, a much more interesting paradox over whom journalists and historians could puzzle. Thus, the legend of Speer as the sole penitent of one of history's most hated regimes was born in 1946—and still survives somewhat today, oddly.

We know what history's verdict is now on Albert Speer, but what will it be in 25, 50, or 100 years? I suspect that then—like some of the Roman emperors we recall now—he, in concert with the late Nikita S. Khrushchev, will be more remembered as the premier chronicler of the system in which he served, than as one of its highest-reigning satraps.

Professor Speer in the final phase of his life—no longer an architect or a prisoner—but as a global best-selling author of several books. In the first, he recalled that being a member of the Nazi Party relieved him of "having to think," but just to do. His Dutch biographer, Dan van der Vat, panned *Spandau: the secret diaries*, and called their author "A lonely man for the last 15 years of his life, always looking backwards… I took away an impression of a patrician old man, arrogant, charming and aloof in equal measure. To this day, I remember much more vividly the taxi driver who brought us together." (LC)

BIBLIOGRAPHY

Andrus, Burton C. *I was the Nuremberg Jailer* (Coward-McCann, 1969); published in the UK under the title *The Infamous of Nuremberg* (London, England: Frewin, 1969)

Davidson, Eugene. *The Trial of the Germans: an account of the twenty-two defendants before the International Military Tribunal at Nuremberg, 1945–46* (New York: Macmillan, 1966)

Dear, I. C. B. (ed.). *The Oxford Companion to the Second World War* (Oxford, England: Oxford University Press, 1995)

Domarus, Max. *Hitler: Reden und Proklamationen, 1932–1945 (Hitler: speeches and proclamations 1932–1945*, tr. M. F. Gilbert (London: Tauris, 1990–)

Dornberger, Walter. *V2: The Nazi Rocket Weapon (V2: des Schuss ins Weltall*, tr. J. Cleugh & G. Halliday (London: Hurst and Blackett, 1954)

Fest, Joachim. *Speer: the Final Verdict*, tr. E. Osers & A. Dring (London, England: Weidenfeld & Nicolson, 2001)

Gaskin, Hilary. *Eyewitnesses at Nuremberg* (London, England: Arms and Armour, 1990)

Goda, Norman J. W. *Tales from Spandau: Nazi Criminals and the Cold War* (New York and Cambridge: Cambridge University Press, 2007)

Gregor, Neil. *Daimler-Benz in the Third Reich* (New Haven and London: Yale University Press, 1998)

Grunberger, Richard. *The 12-Year Reich: A Social History of Nazi Germany, 1933–45* (New York: Holt, Rinehart and Winston, 1971)

Kelley, Douglas M. *22 Cells in Nuremberg* (W. H. Allen, 1947)

King, Benjamin & Kutta, Timothy J. *Impact: The History of Germany's V-Weapons in World War II* (Staplehurst, England: Spellmount, 1998)

Krier, Leon. *Albert Speer. Architecture, 1932–42* (Princeton, NJ: Princeton Architectural Press, 1985)

Larmer, Karl. *Autobahnbau in Deutschland 1933 bis 1945: Zu den Huntergründen (Autobahn Building in Germany, 1933–45)* (Germany: Akademie-Verlag, 1975)

Lepage, Jean-Denis. *The Westwall (Siegfried Line) 1938–45* (The Nafziger Collection, 2002)

Le Tissier, Tony. *The Third Reich Then and Now* (London: Battle of Britain International Ltd, 2005)

Lochner, Louis P. (tr. & ed.). *The Goebbels Diaries* (London: H. Hamilton, 1948)

Neave, Airey. *On Trial at Nuremberg* (London, England: Hodder Stoughton, 1979)

Norden, Eric. "Albert Speer: Playboy interview," *Playboy* (June 1971, 18 (6))

Overy, R. J. *War and Economy in the Third Reich* (Oxford, England: Clarendon Press, 1994)

Patel, Kiran Klaus. *Soldiers of Labor: Labor Service in Nazi Germany and New Deal America, 1933–1945* (Washington, DC: German Historical Institute, 2005)

Persico, Joseph E. *Nuremberg: Infamy on Trial* (New York: Penguin, 1995)

Schmidt, Matthias. *Albert Speer: The End of a Myth* (London, England: Harrap, 1984)

Schönleben, Eduard. *Fritz Todt: Der Mensch, der Ingeniur, der National-sozialist (Fritz Todt: The Man, the Engineer, the National Socialist)* (Oldenburg, Germany: G. Stalling, 1943)

Sereny, Gitta. *Albert Speer: His battle with truth* (London, England: Macmillan, 1995)

Shand, James D. "The Reichsautobahn: Symbol of the Third Reich," *Journal of Contemporary History* (SAGE; Vol. 19, 1984)

Snyder, Louis L. *Encyclopedia of the Third Reich* (New York: McGraw-Hill, 1976)

Speer, Albert. *Inside the Third Reich: Memoirs (Erinnerungen)*, trans. R. & C. Winston (New York and Toronto: Macmillan, 1970)

Speer, Albert. *Spandau: the secret diaries, (Spandauer Tagebücher)* tr. R. & C. Winston (New York: Macmillan, 1976)

Speer, Albert. *Infiltration: How Heinrich Himmler Schemed to Build an SS Industrial Empire (Der Sklavenstaat: meine Auseinandersetzungen mit der SS)*, trans. J. Neugroschel (London. England: Macmillan, 1981)

Spotts, Frederic. *Hitler and the Power of Aesthetics* (London, England: Hutchinson, 2002)

Taylor, Blaine. *Hitler's Headquarters: From Beer Hall to Bunker, 1920–45* (Washington, DC: Potomac Books, 2007)

Taylor, Telford. *The Anatomy of the Nuremberg Trials* (New York: Alfred A. Knopf, 1992)

Todt, Fritz. "Adolf Hitler and His Highways" in *Hitler's Propaganda* (ed. Josef Goebbels) (Munich, Germany: Eher Verlag, 1936)

Tooze, Adam J. *The Wages of Destruction* (London: Allen Lane, 2006)

Van der Vat, Dan. *The Good Nazi: the life and lies of Albert Speer* (London: Weidenfeld & Nicolson, 1997)

Vosselman, Arend. *Reichs-autobahn: Schönheit—Nature—Technik (Reichsautobahn: Beauty, Nature, and Technology)* (Arndt, 2001)

Weinberg, Gerhard L. *Hitler's Foreign Policy, 1933–39: The Road to World War II* (New York: Enigma, 2005)

Wistrich, Robert S. *Who's Who in Nazi Germany* (London, England: Weidenfeld & Nicolson, 1982)

Zaloga, Steven J. *The Atlantic Wall: France* (Oxford, England: Osprey Publishing, 2007)

Zaloga, Steven J. *The Siegfried Line: Battles on the German Frontier, 1944–45* (Oxford, England: Osprey Publishing, 2007)

Christian Zentner & Friedemann Beduerftig (eds.). *Encyclopedia of the Third Reich*, 2 vols, English translation ed. Amy Hackett (New York: Macmillan, 1991)

From left to right: German Army OKW chief Field Marshal Wilhelm Keitel, two unknown officers, Dr. Ferdinand Porsche the Elder, unidentified officer, and Hitler at a weapons demonstration during the war. (Walter Frentz)

ENDNOTES

Chapter One

1 Albert Speer. *Inside the Third Reich: Memoirs*, tr. R. & C. Winston (New York: Macmillan, 1970).

2 Eduard Schönleben. *Fritz Todt: Der Mensch, der Ingeniur, der National-sozialist (Fritz Todt: The Man, the Engineer, the National Socialist)* (Oldenburg: G. Stalling, 1943); Louis L. Snyder. *Encyclopedia of the Third Reich* (New York: McGraw-Hill, 1976).

3 Hitler's eulogy at Fritz Todt's funeral, as recorded by the Foreign Broadcast Monitoring Service, United States Federal Communications Commission, Washington, DC.

4 Schönleben. *Fritz Todt: Der Mensch, der Ingeniur, der National-sozialist.*

5 Hitler's eulogy at Fritz Todt's funeral.

6 Robert S. Wistrich. *Who's Who in Nazi Germany* (London: Weidenfeld & Nicolson, 1982).

7 Speer. *Inside the Third Reich: Memoirs.*

8 Speer. *Inside the Third Reich: Memoirs.*

9 Louis P. Lochner (tr. & ed.). *The Goebbels Diaries* (London: H. Hamilton, 1948).

10 Matthias Schmidt. *Albert Speer: The End of a Myth* (London: Harrap, 1984).

11 Kiran Klaus Patel. *Soldiers of Labor: Labor Service in Nazi Germany and New Deal America, 1933–1945* (Washington, DC: German Historical Institute, 2005).

12 Patel. *Soldiers of Labor.*

Chapter Two

1 "Hitler Takes Austria," *Life magazine* (March 28, 1938).

2 Arend Vosselman. *Reichs-autobahn: Schönheit—Nature—Technik (The Reichsautobahn: Beauty, Nature, and Technology)* (Arndt, 2001).

3 *Deutschlands Autobahnen: Hitlers Strassen (Germany's Highways: Adolf Hitler's Roads)* (General Inspectorate for the German Highway Department, 1937).

4 Vosselman. *Reichs-autobahn.*

5 Fritz Todt. "Adolf Hitler and His Highways" in *Hitler's Propaganda*, ed. Josef Goebbels (Munich: Eher Verlag, 1936).

6 James D. Shand. "The Reichsautobahn: Symbol of the Third Reich," *Journal of Contemporary History* (SAGE; Vol. 19, 1984).

7 Vosselman. *Reichs-autobahn.*

8 Christian Zentner & Friedemann Beduerftig (eds.). *Encyclopedia of the Third Reich*, 2 vols, English translation ed. Amy Hackett (New York: Macmillan, 1991).

9 Snyder, Louis L. *Encyclopedia of the Third Reich* (New York: McGraw-Hill, 1976).

10 Snyder. *Encyclopedia of the Third Reich.*

11 Neil Gregor. *Daimler-Benz in the Third Reich* (New Haven and London: Yale University Press, 1998).

12 Max Domarus. *Hitler: Reden und Proklamationen, 1932–1945* (*Hitler: speeches and proclamations 1932–1945*), trans. M. F. Gilbert (London: Tauris, 1990–); Vol.1.

13 Adam J. Tooze. *The Wages of Destruction* (London: Allen Lane, 2006).

14 Tooze. *The Wages of Destruction.*

15 Tooze. *The Wages of Destruction.*

16 Vosselman. *Reichs-autobahn.*

17 Zentner & Beduerftig. *Encyclopedia of the Third Reich.*

18 Shand. "The Reichsautobahn."

19 Tooze. *The Wages of Destruction.*

20 Vosselman. *Reichs-autobahn.*

21 Vosselman. *Reichs-autobahn.*

22 Todt. "Adolf Hitler and His Highways."

23 Todt. "Adolf Hitler and His Highways."

24 Tony Le Tissier. *The Third Reich Then and Now* (London: Battle of Britain International Ltd, 2005).

25 Le Tissier. *The Third Reich Then and Now.*

26 Frederic Spotts. *Hitler and the Power of Aesthetics* (London: Hutchinson, 2002).

27 *Deutschlands Autobahnen.*

28 Vosselman. *Reichs-autobahn.*

29 Spotts. *Hitler and the Power of Aesthetics.*

30 Vosselman. *Reichs-autobahn.*

31 *Types of German Steel Autobahn Bridges* (US Army Corps of Engineers, 1947).

32 Shand. "The Reichsautobahn."

33 Shand. "The Reichsautobahn."

34 Zentner & Beduerftig. *Encyclopedia of the Third Reich.*

35 Vosselman. *Reichs-autobahn.*

36 *Deutschlands Autobahnen.*

37 *Deutschlands Autobahnen.*

38 *Deutschlands Autobahnen.*

39 Zentner & Beduerftig. *Encyclopedia of the Third Reich.*

40 Karl Larmer. *Autobahnbau in Deutschland 1933 bis 1945: Zu den Huntergründen* (*Autobahn Building in Germany, 1933–45*) (Germany: Akademie-Verlag, 1975).

41 Zentner & Beduerftig. *Encyclopedia of the Third Reich.*

42 Zentner & Beduerftig. *Encyclopedia of the Third Reich.*

43 Zentner & Beduerftig. *Encyclopedia of the Third Reich.*

44 *Deutschlands Autobahnen.*

45 Spotts. *Hitler and the Power of Aesthetics.*

46 Vosselman. *Reichs-autobahn.*

47 Vosselman. *Reichs-autobahn.*

48 Larmer. *Autobahnbau in Deutschland.*

49 Larmer. *Autobahnbau in Deutschland.*

50 Larmer. *Autobahnbau in Deutschland.*

51 Tom Lewis. *Divided Highways: building the interstate highways, transforming American life* (New York: Penguin, 1997).

52 Dan McNichol. *The Roads That Built America: The Incredible Story of the US Interstate System* (Silver Lining Books, 2003).

53 Dwight D. Eisenhower. *The White House Years, Vol. 1: Mandate for Change 1953–56* (Garden City, NY: Doubleday, 1963).

54 McNichol. *The Roads That Built America.*

55 McNichol. *The Roads That Built America.*

Chapter Three

1 Telford Taylor. *Munich: The Price of Peace* (Garden City, NY: Doubleday, 1979).

2 Jean-Denis Lepage. *The Westwall (Siegfried Line) 1938–45* (The Nafziger Collection. 2002).

3 Taylor. *Munich.*

4 Taylor. *Munich.*

5 Taylor. *Munich.*

6 Lepage. *The Westwall.*

7 Lepage. *The Westwall.*

8 Tooze. *The Wages of Destruction.*

9 Lepage. *The Westwall.*

10 Gerhard Engel. *At the Heart of the Reich: the secret diary of Hitler's Army Adjutant* (trans. G. Brooks) (London: Greenhill Books, 2005), August 14, 1938.

11 Lepage. *The Westwall.*

12 Zentner & Beduerftig. *Encyclopedia of the Third Reich.*

13 Domarus. *Hitler: Reden und Proklamationen,* Vol. 3.

14 Domarus. *Hitler: Reden und Proklamationen,* Vol. 3.

15 Taylor. *Munich.*

16 Snyder. *Encyclopedia of the Third Reich.*

17 Zentner & Beduerftig. *Encyclopedia of the Third Reich.*

18 Blaine Taylor. *Hitler's Headquarters: From Beer Hall to Bunker, 1920–45* (Washington, DC: Potomac Books, 2007).

19 Rolf-Dieter Müller. "Todt Organization," *The Oxford Companion to World War II*, ed. I. C. B. Dear & M. R. D. Foot (Oxford: Oxford University Press, 2001).

20 Müller. "Todt Organization."

21 Taylor. *Munich*.

22 Lepage. *The Westwall*.

23 Taylor. *Munich*.

24 Domarus. *Hitler: Reden und Proklamationen*, Vol. 3.

25 Domarus. *Hitler: Reden und Proklamationen*, Vol. 3.

26 Lepage. *The Westwall*.

27 *Dr. Todt: Berufung und Werk (Todt: Man, Mission and Achievement)* (1943).

28 General Siegfried Westphal. *The German Army in the West* (London, 1951).

29 "Maginot Line," *Encyclopedia Brittanica* (London: Encyclopedia Brittanica, Inc., 1966).

30 Domarus. *Hitler: Reden und Proklamationen*, Vol. 3.

31 George S. Patton's posthumously published memoirs: *War as I Knew It* (Houghton Mifflin, 1947).

32 Steven J. Zaloga. *The Siegfried Line: Battles on the German Frontier, 1944–45* (Oxford, England: Osprey Publishing, 2007).

33 Zaloga. *The Siegfried Line*.

34 Omar Nelson Bradley & Clay Blair. *A General's Life: an autobiography* (London: Sidgwick & Jackson, 1983).

35 Bradley. *A General's Life*.

36 Bradley. *A General's Life*.

37 Bradley. *A General's Life*.

38 Bradley. *A General's Life*.

39 Lepage. *The Westwall*.

Chapter Four

1 Speech by Adolf Hitler, 1935, quoted in Domarus. *Hitler: Reden und Proklamationen*, Vol. 2.

2 Gerhard L. Weinberg. *Hitler's Foreign Policy, 1933–39: The Road to World War II* (New York: Enigma, 2005).

3 Schönleben. *Fritz Todt: Der Mensch, der Ingeniur, der National-sozialist*.

4 Tooze. *The Wages of Destruction*.

5 Tooze. *The Wages of Destruction*.

6 Tooze. *The Wages of Destruction*.

7 Tooze. *The Wages of Destruction*.

8 Tooze. *The Wages of Destruction*.

9 Tooze. *The Wages of Destruction*.

10 Tooze. *The Wages of Destruction.*

11 Dan van der Vat. *The Good Nazi: the life and lies of Albert Speer* (London: Weidenfeld & Nicolson, 1997).

12 Tooze. *The Wages of Destruction.*

13 Domarus. *Hitler: Reden und Proklamationen,* Vol. 3.

14 Tooze. *The Wages of Destruction.*

15 Tooze. *The Wages of Destruction.*

16 Tooze. *The Wages of Destruction.*

17 Tooze. *The Wages of Destruction.*

18 Van der Vat. *The Good Nazi.*

19 Tooze. *The Wages of Destruction.*

20 Speer, *Inside the Third Reich: Memoirs.*

21 Tooze. *The Wages of Destruction.*

22 Tooze. *The Wages of Destruction.*

23 Van der Vat. *The Good Nazi.*

24 Tooze. *The Wages of Destruction.*

25 "Engineering: A General Survey," *Grolier Encyclopedia* (Danbury, CT: Grolier, 1946).

26 "Engineering: A General Survey."

27 "Engineering: A General Survey."

28 I. C. B. Dear (ed.). *The Oxford Companion to the Second World War* (Oxford: Oxford University Press, 1995).

29 Dear. *The Oxford Companion to the Second World War.*

30 Dear. *The Oxford Companion to the Second World War.*

31 Tooze. *The Wages of Destruction.*

32 Eric Norden. "Albert Speer: *Playboy* interview" *Playboy magazine* (June 1971, Vol.18/6).

33 Tooze. *The Wages of Destruction.*

34 Tooze. *The Wages of Destruction.*

35 Tooze. *The Wages of Destruction.*

36 Schmidt. *Albert Speer: The End of a Myth.*

37 Van der Vat. *The Good Nazi.*

38 Tooze. *The Wages of Destruction.*

39 Norden. "Albert Speer: *Playboy* interview."

40 Norden. "Albert Speer: *Playboy* interview."

Chapter Five

1 Lochner. *The Goebbels Diaries.*

2 Lochner. *The Goebbels Diaries.*

3 Lochner. *The Goebbels Diaries.*

4 Lochner. *The Goebbels Diaries.*

5 Hitler's eulogy at Fritz Todt's funeral.

6 Leon Krier. *Albert Speer. Architecture, 1932–42* (Princeton, NJ: Princeton Architectural Press, 1985).

7 Lochner. *The Goebbels Diaries.*

8 Schmidt. *Albert Speer: The End of a Myth.*

9 Schmidt. *Albert Speer: The End of a Myth.*

10 Speer. *Inside the Third Reich: Memoirs.*

11 Speer. *Inside the Third Reich: Memoirs.*

12 Accident report [K 1 T.L. 11/42], quoted in Speer. *Spandau: The Secret Diaries.*

13 Speer. *Inside the Third Reich: Memoirs.*

14 Schmidt. *Albert Speer: The End of a Myth*

15 Schmidt. *Albert Speer: The End of a Myth*

16 Hans Baur. *Hitler's Pilot (Ich flog mächtige der Erde)* (trans. E. Fitzgerald) (London: Müller, 1958); republished as Hans Baur (1986). *Hitler at My Side* (trans. L. Butler) (Houston TX: Eichler Pub. Corp.).

17 Speer. *Spandau: The Secret Diaries.*

18 Baur. *Hitler at My Side.*

19 Speer. *Inside the Third Reich: Memoirs.*

20 Dr. Speer himself noted in his book: Speer. *Inside the Third Reich: Memoirs.*

21 Speer. *Inside the Third Reich: Memoirs.*

22 Speer. *Inside the Third Reich: Memoirs.*

23 Gitta Sereny. *Albert Speer: His battle with truth* (London: Macmillan, 1995).

24 Schmidt. *Albert Speer: The End of a Myth*, which was subtitled, "The book that exposes Albert Speer's falsification of history and his real role in the Third Reich."

25 Schmidt. *Albert Speer: The End of a Myth.*

26 *Schmidt. Albert Speer: The End of a Myth.*

Chapter Six

1 Van der Vat. *The Good Nazi.*

2 Sereny. *Albert Speer: His battle with truth.*

3 Michael Getler. "Albert Speer: An Invaluable Link to Understanding Nazi Germany," *The Washington Post* (February 23, 1976)

4 Joachim Fest. *Speer: the Final Verdict* (trans. E. Osers & A. Dring) (London: Weidenfeld & Nicolson, 2001).

5 Speer, *Inside the Third Reich: Memoirs.*

6 Steven J. Zaloga. *The Atlantic Wall: France* (Oxford: Osprey Publishing, 2007).

7 Speer. *Inside the Third Reich: Memoirs*.

8 Blaine Taylor. "Health in History/German Armaments Minister Albert Speer," *Maryland State Medical Journal* (February 1978).

9 Taylor. "Health in History."

10 Taylor. "Health in History."

11 Omer Bartov. "The Mask of Decency," review of Joachim Fest, *Speer: The Final Verdict*, *The New Republic* (November 18, 2002): 36–41.

12 Phyllis G. Proctor. "Germany's miracle-working armaments czar used slave labor 'without knowing' the full facts," *World War II Magazine*.

13 Fest. *Speer: the Final Verdict*.

14 Speer. *Inside the Third Reich: Memoirs*.

15 Fest. *Speer: the Final Verdict*.

16 Richard Suchenwirth. *Command and Leadership in the German Air Force*, USAF historical studies no.174 (New York: Arno Press, 1969).

17 Fest. *Speer: the Final Verdict*.

18 Fest. *Speer: the Final Verdict*.

19 Fest. *Speer: the Final Verdict*.

20 Speer. *Inside the Third Reich: Memoirs*.

Chapter Seven

1 Walter Dornberger. *V2: The Nazi Rocket Weapon (V2: des Schuss ins Weltall)* (trans. J. Cleugh & G. Halliday) (London: Hurst and Blackett, 1954).

2 Benjamin King & Timothy J. Kutta. *Impact: The History of Germany's V-Weapons in World War II* (Staplehurst: Spellmount, 1998).

3 Dornberger. *V2: The Nazi Rocket Weapon*.

4 Dornberger. *V2: The Nazi Rocket Weapon*.

5 Thomas O'Toole. "W. R. Dornberger dies, German Rocket Expert," *The Washington Post* (July 2, 1980).

6 Dornberger. *V2: The Nazi Rocket Weapon*.

7 Dornberger. *V2: The Nazi Rocket Weapon*.

8 Speer. *Inside the Third Reich: Memoirs*.

9 Dornberger. *V2: The Nazi Rocket Weapon*.

10 Speer. *Inside the Third Reich: Memoirs*.

11 Dornberger. *V2: The Nazi Rocket Weapon*.

12 Dornberger. *V2: The Nazi Rocket Weapon*.

13 Dornberger. *V2: The Nazi Rocket Weapon*.

14 According to David M. Glantz (ed.). *Hitler and His Generals: Military Conferences, 1942–1945* (London: Greenhill, 2002).

15 Sereny. *Albert Speer: His battle with truth.*

16 Speer. *Infiltration.*

17 Dornberger. *V2: The Nazi Rocket Weapon.*

18 Dornberger. *V2: The Nazi Rocket Weapon.*

19 Dornberger. *V2: The Nazi Rocket Weapon.*

20 Dornberger. *V2: The Nazi Rocket Weapon.*

21 Dornberger. *V2: The Nazi Rocket Weapon.*

22 Dornberger. *V2: The Nazi Rocket Weapon*

23 Dornberger. *V2: The Nazi Rocket Weapon.*

24 King & Kutta. *Impact.*

25 Dwight D. Eisenhower. *Crusade in Europe* (New York: Doubleday, 1948).

26 Albert Speer. *Spandau: the secret diaries*, tr. R. & C. Winston (New York: Macmillan, 1976); March 18, 1947.

27 Dornberger. *V2: The Nazi Rocket Weapon.*

28 Dornberger. *V2: The Nazi Rocket Weapon.*

29 Dornberger. *V2: The Nazi Rocket Weapon.*

30 M. G. Lord. "Review of Michael J. Neufeld. *Von Braun: Dreamer of Space, Engineer of War*," *Baltimore Sun* (September 30, 2007).

31 Zentner & Beduerftig. *Encyclopedia of the Third Reich.*

32 Jimmy Carter. Obituary, Wernher von Braun, *Time* (June 27, 1977).

33 Albert Speer. *Infiltration: How Heinrich Himmler Schemed to Build an SS Industrial Empire (Der Sklavenstaat: meine Auseinandersetzungen mit der SS)*, tr. J. Neugroschel (London: Macmillan, 1981).

34 Speer. *Infiltration.*

35 Dornberger. *V2: The Nazi Rocket Weapon.*

36 Dornberger. *V2: The Nazi Rocket Weapon.*

37 Dornberger. *V2: The Nazi Rocket Weapon.*

38 Taylor. *Hitler's Headquarters.*

39 Sereny. *Albert Speer: His battle with truth.*

40 Sereny. *Albert Speer: His battle with truth.*

41 Speer. *Infiltration.*

42 Speer. *Infiltration.*

43 Sereny. *Albert Speer: His battle with truth.*

44 Sereny. *Albert Speer: His battle with truth.*

45 Emily Newburger (2002). "Never Forget," *Harvard Law Review*; "Former Nazi scientist to return to Germany," *Baltimore* (Sun, August 11, 1990).

46 Dornberger. *V2: The Nazi Rocket Weapon.*

Chapter Eight

1 John F. Crossland, "The Secret Nazi Atomic Tapes," *The Washington Post* (March 1, 1992).

2 Crossland. "The Secret Nazi Atomic Tapes."

3 Crossland. "The Secret Nazi Atomic Tapes."

4 Crossland. "The Secret Nazi Atomic Tapes."

5 Crossland. "The Secret Nazi Atomic Tapes."

6 Crossland. "The Secret Nazi Atomic Tapes."

7 Crossland. "The Secret Nazi Atomic Tapes."

8 Speer. *Inside the Third Reich: Memoirs.*

9 Speer. *Inside the Third Reich: Memoirs.*

10 Speer. *Inside the Third Reich: Memoirs.*

11 Speer. *Inside the Third Reich: Memoirs.*

12 Speer. *Inside the Third Reich: Memoirs.*

13 Sereny. *Albert Speer: His battle with truth.*

14 Sereny. *Albert Speer: His battle with truth.*

15 Sereny. *Albert Speer: His battle with truth.*

16 Sereny. *Albert Speer: His battle with truth.*

17 Tooze. *The Wages of Destruction.*

Chapter Nine

1 Lochner. *The Goebbels Diaries.*

2 Speer. *Inside the Third Reich: Memoirs.*

3 David Edgar. *Albert Speer* (London: Nick Hern Books, 2000).

4 Sereny. *Albert Speer: His battle with truth.*

5 Speer. *Infiltration.*

6 Speer. *Inside the Third Reich: Memoirs.*

7 Speer. *Inside the Third Reich: Memoirs.*

8 Speer. *Inside the Third Reich: Memoirs.*

9 Speer. *Inside the Third Reich: Memoirs.*

10 Speer. *Inside the Third Reich: Memoirs.*

11 Speer. *Inside the Third Reich: Memoirs.*

12 R. J. Overy. *War and Economy in the Third Reich* (Oxford: Clarendon Press, 1994).

13 Speer. *Inside the Third Reich: Memoirs.*

14 Speer. *Inside the Third Reich: Memoirs.*

15 Speer. *Inside the Third Reich: Memoirs.*

16 Tooze. *The Wages of Destruction.*

17 Overy. *War and Economy in the Third Reich.*

18 Overy. *War and Economy in the Third Reich.*

19 Tooze. *The Wages of Destruction.*

20 Overy. *War and Economy in the Third Reich.*

21 Speer. *Inside the Third Reich: Memoirs.*

22 Speer. *Inside the Third Reich: Memoirs.*

23 Tooze. *The Wages of Destruction.*

24 Overy. *War and Economy in the Third Reich.*

25 Overy. *War and Economy in the Third Reich.*

26 Overy. *War and Economy in the Third Reich.*

27 Overy. *War and Economy in the Third Reich.*

28 Overy. *War and Economy in the Third Reich.*

29 Schmidt. *Albert Speer: The End of a Myth.*

30 Sereny. *Albert Speer: His Battle with truth.*

31 Sereny. *Albert Speer: His Battle with truth.*

32 Norden. "Albert Speer: *Playboy* interview."

33 Overy. *War and Economy in the Third Reich.*

34 Grunberger, Richard. *The 12-Year Reich: A Social History of Nazi Germany, 1933–45* (New York: Holt, Rinehart and Winston, 1971).

35 Norden. "Albert Speer: *Playboy* interview."

36 Norden. "Albert Speer: *Playboy* interview."

37 Tooze. *The Wages of Destruction.*

38 Speer. *Spandau.*

39 Speer. *Inside the Third Reich: Memoirs.*

40 Sereny. *Albert Speer: His battle with truth.*

41 Norden. "Albert Speer: Playboy interview."

42 Speer. *Inside the Third Reich: Memoirs.*

43 Speer. *Inside the Third Reich: Memoirs.*

44 Sereny. *Albert Speer: His battle with truth.*

45 Speer. *Inside the Third Reich: Memoirs.*

46 Lochner. *The Goebbels diaries.*

47 Speer. *Inside the Third Reich: Memoirs.*

48 Sereny. *Albert Speer: His battle with truth.*

49 Norden. "Albert Speer: *Playboy* interview."

50 Van der Vat. *The Good Nazi.*

51 Fest. *Speer.*

52 Speer. *Inside the Third Reich: Memoirs.*

53 Speer. *Infiltration.*

54 Sereny. *Albert Speer: His battle with truth.*

55 Speer. *Inside the Third Reich: Memoirs.*

56 Speer. *Inside the Third Reich: Memoirs.*

57 Speer. *Inside the Third Reich: Memoirs.*

58 Sereny. *Albert Speer: His battle with truth.*

59 Sereny. *Albert Speer: His battle with truth.*

60 Speer. *Inside the Third Reich: Memoirs.*

61 Speer. *Inside the Third Reich: Memoirs.*

62 Speer. *Inside the Third Reich: Memoirs.*

63 Sereny. *Albert Speer: His battle with truth.*

64 Norden. "Albert Speer: *Playboy* interview."

65 Speer. *Infiltration.*

66 Speer. *Inside the Third Reich: Memoirs.*

67 Sereny. *Albert Speer: His battle with truth.*

68 Norden. "Albert Speer: *Playboy* interview."

69 Speer. *Infiltration.*

70 Speer. *Infiltration.*

71 Sereny. *Albert Speer: His battle with truth.*

72 Speer. *Inside the Third Reich: Memoirs.*

73 Speer. *Inside the Third Reich: Memoirs.*

74 Speer. *Inside the Third Reich: Memoirs.*

75 Sereny. *Albert Speer: His battle with truth.*

76 Speer. *Inside the Third Reich: Memoirs.*

77 Sereny. *Albert Speer: His battle with truth.*

78 Sereny. *Albert Speer: His battle with truth.*

79 Speer. *Inside the Third Reich: Memoirs.*

80 John Kenneth Galbraith and George W. Ball. "The Interrogation of Albert Speer," *Life* (December 17, 1945)

81 Sereny. *Albert Speer: His battle with truth.*

82 Sereny. *Albert Speer: His battle with truth.*

83 Galbraith & Ball. "The Interrogation of Albert Speer."

Chapter Ten

1 Speer. *Inside the Third Reich: Memoirs.*

2 Karel Margry. "The Flensburg Government," *After the Battle* (Issue 128).

3 Speer. *Inside the Third Reich: Memoirs.*

4 Speer. *Inside the Third Reich: Memoirs.*

5 Galbraith & Ball. "The Interrogation of Albert Speer."

6 Galbraith & Ball. "The Interrogation of Albert Speer."

7 Galbraith & Ball. "The Interrogation of Albert Speer."

8 Margry. "The Flensburg Government."

9 Speer. *Inside the Third Reich: Memoirs.*

10 Margry. "The Flensburg Government."

11 Speer. *Inside the Third Reich: Memoirs.*

12 Margry. "The Flensburg Government."

13 Speer. *Inside the Third Reich: Memoirs.*

14 Speer. *Inside the Third Reich: Memoirs.*

15 Speer. *Inside the Third Reich: Memoirs.*

16 Speer. Inside the Third Reich: Memoirs.

17 Speer. *Inside the Third Reich: Memoirs.*

18 Airey Neave. *On Trial at Nuremberg* (London: Hodder Stoughton, 1979).

19 Neave. *On Trial at Nuremberg.*

20 Neave. *On Trial at Nuremberg.*

21 Burton C. Andrus. *I was the Nuremberg Jailer* (Coward-McCann, 1969).

22 Andrus. *I was the Nuremberg Jailer.*

23 Sereny. *Albert Speer: His battle with truth.*

24 Douglas M. Kelley. *22 Cells In Nuremberg* (W. H. Allen, 1947).

25 Speer. *Spandau.*

26 John K. Lattimer. *Hitler's Fatal Sickness and Other Secrets of the Nazi Leaders* (Hippocrene Books, 1999), p.246.

27 Drexel A. Sprecher. "Letters to the Editor," *The Washington Post* (September 8, 1981).

28 Taylor also assessed Speer in his book: Telford Taylor. *The Anatomy of the Nuremberg Trials* (New York: Alfred A. Knopf, 1992).

29 Bradley F. Smith. *Reaching Judgment at Nuremberg* (New York: Basic Books, 1977).

30 Smith. *Reaching Judgment at Nuremberg.*

31 Norman J. W. Goda. *Tales from Spandau: Nazi Criminals and the Cold War* (New York and Cambridge: Cambridge University Press, 2007).

31 Joseph E. Persico. *Nuremberg: Infamy on Trial* (New York: Penguin, 1995).

33 Hilary Gaskin. *Eyewitnesses at Nuremberg* (London: Arms and Armour, 1990).

34 Eugene Davidson. *The Trial of the Germans: an account of the twenty-two defendants before the International Military Tribunal at Nuremberg, 1945–46* (New York: Macmillan, 1966).

35 Davidson. *The Trial of the Germans.*

Chapter Eleven

1 Speer. *Spandau.*

2 Tony Le Tissier. *Berlin Then and Now* (Old Harlow, Essex: After the Battle, 1992).

3 Jack Fishman. *The Seven Men of Spandau* (Rinehart, 1954).

4 Speer. *Spandau.*

5 Speer. *Spandau.*

6 Speer. *Spandau.*

7 Speer. *Spandau.*

8 Speer. *Spandau.*

9 Speer. *Spandau.*

10 Goda, Norman J. W. *Tales from Spandau: Nazi Criminals and the Cold War* (New York; Cambridge: Cambridge University Press, 2007).

11 Sereny. *Albert Speer: His battle with truth.*

12 Speer. *Spandau.*

13 Speer. *Spandau.*

14 Speer. *Spandau.*

15 Speer. *Spandau.*

16 Speer. *Spandau.*

17 Speer. *Spandau.*

18 Speer. *Spandau.*

19 Speer. *Spandau.*

20 Speer. *Spandau.*

21 Speer. *Spandau.*

22 Speer. *Spandau.*

23 Speer. *Spandau.*

24 Speer. *Spandau.*

25 Speer. *Spandau.*

26 Speer. *Spandau.*

27 Speer. *Spandau.*

28 Sprecher. "Letters to the Editor."

29 Schmidt. *Albert Speer: The End of a Myth.*

30 Schmidt. *Albert Speer: The End of a Myth.*

31 "Albert Speer, 76, Dies; Director of Hitler's Industrial Machine," *The Washington Post* (September 2, 1981).

32 "Albert Speer dies at 76; Close Associate of Hitler," *The New York Times* (September 2, 1981).